NATIONAL PLANNING
A S S O C I A T I O N

D1255405

SEEDS
AND WORLD
AGRICULTURAL
PROGRESS

Neil McMullen

SB 114
A3
M3
1987

**Seeds and World Agricultural
Progress**

NPA Report #227

Price: $25.00

ISBN 0-89068-088-4
Library of Congress
Catalog Card Number 87-60766

Printed in the United States of America

To my son
Robert Wardale McMullen
January 22, 1974 – October 28, 1986

Rob accompanied me on part of the
field work for this book
and helped me in many ways.

Contents

Figures

Seeds and the World Food Balance

1

KEY DETERMINANTS OF THE WORLD FOOD BALANCE

To put the production and trade of seeds into proper focus, a brief overview of the world food balance is helpful. This balance is multi-dimensional, constantly evolving and characterized by seeming contradictions. These contradictions and much of the ongoing change in the world food balance can be explained by the interactions of the following six determinants:

    (1) long-term population growth patterns and prospects;
    (2) changes in per capita food consumption;
    (3) medium-term food supply disturbances;
    (4) nutritional and food distribution problems;
    (5) government policies affecting agriculture and food;
    (6) the production performance of the agricultural sector over time.

Long-term Population Growth

Until 1750, population growth was very slow and characterized by instability and periodic declines. During the fourteenth century, for example, European population declined sharply, with estimates of the decline between 1346 and 1400 as high as 50 percent.[1] This decline, attributable to the bubonic plague, was particularly dramatic, but population trends before 1750 were characterized by fluctuations rather than continuing secular growth. Beginning in England in the 1740s and in other parts of Europe in subsequent decades, however, sustained growth raised the population in Great Britain from 8.9 million in 1780 to 37.1 million in 1900[2] and in France from 27.5 million in 1800 to 38.8 million in 1900. In Germany, growth came later, when the population grew from 41.1 million in 1871 to 56.4 million in 1900 and 64.9 million in 1910.[3] In addition to this increase, millions emigrated to America and other newly settled lands. In Western Europe, population growth moderated before World War I,[4] and during the second half of the twentieth century, European population growth has been extremely low.

    The dynamic at work is a disequilibrium caused by significant differences in birth and death rates. In preindustrial societies, both birth and death rates were high—about 40 per 1,000—with death rates also marked by instability, resulting in the slow growth marked by periodic declines just noted. At a certain point in a society's development, the combination of better diets and improved health practices began to reduce deaths, particularly childhood deaths, and the annual death rate fell over time, in some cases to levels slightly below 10 per 1,000. However, birth rates tended to remain at the original level before declining at a much slower pace. Equilibrium is restored when the birth rate approximately matches the death rate.

The lag in the birth rate adjustment is due to several factors. First, it can take two generations before it is evident that the majority of children born have lived to adulthood. Therefore, three births in a family virtually guarantees that two children will reach maturity. Second, in agrarian societies, additional children mean additional workers within a few years, and large families per se are viewed as beneficial to society. In all settings, the educational level of women is also an important determinant of the birth rate; as literacy spreads, educational levels rise, opportunities open up for women and birth rates tend to decline.

In Western Europe the "population explosion" was slowing by 1900; in Eastern Europe rapid growth persisted up to World War II. In both cases it took about 100 years for the birth rate to move down to its current equilibrium with the death rate.

The beginning of the rise in European population occasioned the work of Malthus, who pointed out the potential for populations to quickly outrun food supplies, resulting in large-scale famine. Writing in 1798, Malthus warned that population grows geometrically, while food increases arithmetically. However, these dire predictions did not come to pass in Europe during the century and a half of rapid growth that followed. In fact, food supplies stayed ahead of population growth —new lands were brought under cultivation in America and Australia; railroads and steamships made new food supplies available to European markets; and advances in agriculture in Europe increased yields on old farmland. Measured in real terms, i.e., hours worked to purchase a basic diet, food costs fell throughout the period, and Western Europeans were able to improve their diet even as the number of people doubled during the nineteenth century.

Since 1950 the phenomenon of a "population explosion" has shifted to the developing world. From 1950 to 1975 the death rate fell sharply. This was primarily due to improved public health practices, but also to the alleviation of local famines through access to world food stocks on a concessionary basis. As in Europe in the previous century, the dramatic change was the high proportion of newly born who eventually became adults. During this initial period, the proportion of young people 15 years and under rose as infant and child mortality fell.[5]

Most developing countries are still experiencing rapid growth, with a high percentage of youth in the total population. Death rates are typically 1.0 to 2.0 per 100 while birth rates are in the range of 2.5 to 4.0; the difference is the rate of growth of total population.

Compared with the European experience in the 1800s, the death rate in most developing countries has declined rapidly. In Europe it fell gradually and continually for more than 150 years to its current level of less than 10 per 1,000. In the developing world the death rate dropped abruptly, in most cases taking only about 35 years to reach 15 per 1,000.[6] Birth rates have also begun to fall in many developing countries, particularly in small countries with growing economies and relatively high literacy rates. However, birth rates remain around 25 to 40 per 1,000, implying a population growth rate of 2.0 to 2.5 percent per year for most developing countries. In Africa, birth and death rates are typically higher, and the underlying population growth rate is about 3.5 percent per year.

Just as death rates in the developing countries have fallen more rapidly than historical precedents would have predicted, birth rates may not lag for as long as occurred in Europe in the nineteenth century. Whereas a new population equilibrium was achieved in Europe in about 100 years, the process may be accelerated in the developing world. This is already occurring in countries with the most dynamic economies, such as Taiwan, Singapore, South Korea, and Hong Kong, and in the two largest countries, China and India.

The trend of the birth rate in the developing world will be one of the most important determinants of the world food balance in the years ahead. If birth rates decline and move into line with death rates over the next 25 or 30 years, then the population explosion in the developing world will have run its course in 50 to 60 years. Such an outcome implies increasing population pressure on world food supplies into the first decade or two of the next century, but an easing of long-term pressure. On the other hand, a continuation of high birth rates until 2050 or later would put extreme pressure on world food supplies well into the next century and could have an important destabilizing effect on international political and economic systems. In this case, birth rates would stay in the range of 25 to 40 per 1,000, and developing country populations would continue to grow by 2.0 to 2.5 percent per year. The consequences of this rate of growth would be a doubling of population every 30 years.

One intermediate outcome might result from a division of the developing world into countries that are already experiencing rapid economic growth and slow population growth and those that are continuing to expand their populations rapidly and may or may not be growing as fast economically. In the first case, income per capita is definitely rising, whereas in the second case, income per capita may

be stagnating or falling. Thus, the birth rate in the developing countries is probably one of the best indicators of the direction of long-
term pressure not only on the world food balance, but on international
political and economic conditions as well.

The implications of the long-term population trends for world agriculture are unmistakable. Under the best assumption, world population will increase from approximately 4.5 billion in 1980 to 7.0 billion
by 2010.[7] This projection anticipates a gradual decline in the birth rate
between 1980 and 2000, with overall population growth reaching a
low and steady rate by 2020. This implies the need for rapid and continued expansion of world food supplies to meet population growth
for about the next 40 years. During this period, world population growth
would slow to a rate of less than 1 percent, significantly reducing pressure on the world food balance.

In contrast, if birth rates in most developing countries remain high
until 2050 or later, world population growth will be much more rapid,
reaching as high as 15 billion by 2100. Pressure on food supplies would
thus be very strong, and there would be every reason for concern
regarding the availability of food for so many people.

It is clear that world population growth will certainly be a factor
increasing the need for agricultural output for at least the next 40 years,
and perhaps beyond. Only after this period will there be real hope
for low growth in overall population, and this will be possible only
if developing countries continue to moderate their birth rates. Considering the first of the six determinants of the world food balance,
then, long-term population prospects are for continued growth and
continued increases in demand pressure on world food supplies into
at least the first two decades of the next century.

Changes in Per Capita Food Consumption

Per capita income growth is the second key determinant that must
be factored into consideration of the world food balance. As economic
development proceeds and income growth surpasses population growth,
per capita income rises above subsistence levels and people's diets
begin to change and improve. Families and individuals at a subsistence
level of income tend to spend high proportions of any additional income on food. Their income elasticity of demand (percentage increase
in food consumption relative to the percentage increase in income)
is rather high, on the order of 0.9.[8] Thus, if a family spends 72 percent
of its income on food initially and its income elasticity of demand
is 0.9, then it will spend 65 percent of additional income on food. This
figure falls as per capita income rises and, for the more affluent, ap-
proximates 0.1. These patterns prevail both among countries and within
countries. Whenever income grows for poor countries or for poor

households, a large proportion of that income will go to expand and improve diets. Similarly, where income growth occurs in affluent countries or affluent households in poor countries, the proportionate increase in demand for food is small.

Population growth and rises in per capita income seem to relate in a particular pattern to a society's stage of economic development. Population increases most rapidly during the initial stages of development, while per capita income does not begin to rise rapidly until the intermediate stages. The effect of the combination of population and income on the demand for food is shown in Table 1-1. "Low income countries" derive most of their increased demand for food from population growth. The "rapid growth countries," specifically many Latin American countries in the 1970s, have fast population growth and rising per capita income. Their growth in demand for food is derived about evenly from population increases and per capita income growth. Programs of structural adjustment in many of these countries have raised prices of agricultural goods for consumers and producers. Although it is still too soon to draw firm conclusions, the result seems to be reduced consumption locally and increased exports. "Intermediate countries" are characterized by rapid economic growth and declining population growth; examples include several of the newly industrializing countries of Asia, such as South Korea, Taiwan, Singapore, and Hong Kong. At this point, the growth in demand for food is based primarily on increasing per capita income. Finally, "mature countries" have low population growth, moderate per capita income growth rates and low income elasticities of demand for food.

Per capita income clearly is an important factor in determining world food balance. The slowdown in the world economy that occurred in the mid-1970s and more dramatically in the early 1980s has reduced growth in demand for agricultural products. However, if income growth accelerates, then rising per capita incomes will again be an important contributor to increasing demand for food. Assuming that the world economy continues to expand at moderate rates through the 1990s, population growth and per capita income will continue to put demand-side pressures on the world food balance.

An important corollary is the financing of increased food demand. Low income countries generally lack foreign exchange to buy food and therefore depend on local production and concessionary supplies. Many rapid growth countries are currently faced with high foreign indebtedness, an extreme shortage of foreign exchange, and a very limited capacity to import agricultural goods. In these circumstances, domestic food production is the answer, which would also help to absorb the labor of their still rapidly growing populations. The intermediate countries, in contrast, have slower population growth, less suitable land for agriculture, a high level of labor productivity in industry,

TABLE 1-1. LEVELS OF DEVELOPMENT AND DEMAND FOR FOOD

	Low Income Countries	Rapid Growth Countries	Intermediate Countries	Mature Countries
Population growth	2.5	2.5	1.8	0.5
Income growth	4.0	6.0	6.3	4.0
Income per capita growth	1.5	3.5	4.5	3.5
Income elasticity	0.9 to 0.7	0.7 to 0.5	0.7 to 0.5	0.4 to 0.1
Income related growth in demand	1.3	2.0	2.7	0.7
Population related growth in demand	2.5	2.5	1.8	0.5
Total growth in demand	3.8	4.5	4.5	1.2

Sources: Based on Table 7 in *Problems and Prospects for U.S. Agriculture in World Markets*, Timothy Josling (Washington, D.C.: NPA, Committee on Changing International Realities, 1981), p. 16. Income and population trends from World Bank, *World Development Report* (WDR), 1983.

and generally fewer constraints in terms of foreign exchange. For them, the correct course is probably to import more foodstuffs and to focus on exporting high value added industrial goods. This, of course, assumes that world markets remain open to these goods and that protectionism does not severely limit exports. World income growth in the 1980s will therefore affect the world food balance not only in terms of per capita income growth, but also by making foreign exchange more available to potential importers of foodstuffs and by reducing protectionist pressures to restrict international trade. Slow growth in the world economy will reduce demand for food and channel it to domestic markets. Rapid income growth will put more pressure on the overall food balance by raising incomes and therefore demand, while providing countries with the foreign exchange necessary to buy in international markets. While population growth at levels similar to the recent past seems reasonably certain over the next decade, income growth could accelerate, putting increased demand pressure on world food supplies.

Medium-term Disturbances in World Food Supplies

Since the end of World War II, events have periodically created cycles of optimism and pessimism regarding an adequate world supply of foodstuffs. Throughout the 1950s and 1960s, actual levels of agricultural production were less important than the substantial stocks of agricultural products, primarily cereals, held by the United States and a few other food surplus countries. These stocks dominated thinking on the world food balance until the early 1960s. By then surpluses had begun to decline, agricultural prices were trending slowly upward, and a full awareness of the population explosion in the developing countries began strongly to influence views of the world food balance. At the end of the 1960s, this pessimism began to reverse, surplus stocks rose and prices trended down after 1967. Most important, the "Green Revolution" and increasing supplies of fishmeal seemed to indicate good prospects for the adequacy of food supplies for a decade or two. Therefore, as the 1970s began, the world food balance was viewed with increasing optimism, and the issue of how to cope with food surpluses emerged.

In fact, world grain stocks peaked in 1970 and by 1972 had declined by more than 25 percent. Stocks of other foodstuffs declined by smaller amounts, but several additional shocks rapidly followed. After record catches in the late 1960s and good catches up to mid-1972, Peru's anchovy catch fell off disastrously. This was one of the world's more important protein sources for supplementing animal feed, and its failure had profound effects on soybeans, the nearest substitute.

Soybean prices rose fourfold and soybean stocks fell immediately. The second blow was a failure in Russian wheat; massive Russian purchases drove up international prices of wheat, corn and rice and reduced stockpiles. Poor crop yields in several other countries, including India, Bangladesh and China, added to the growing demand for international food supplies. Finally, the steep rise in oil prices pushed up costs of fuel and fertilizer and seemed to ensure that food prices would stay at historically high levels for several years.

During the second half of the 1970s, a new order prevailed in world agricultural markets. Russia continued to import significant quantities of grain from the large exporting countries of Argentina, Australia, Canada, and the United States. Stocks bottomed out in 1976 and began to recover slowly, while prices peaked in 1974 and fell thereafter. It is crucial, however, to note that, while declining in the mid-1970s, agricultural prices rose in the late 1970s and remained at historically high levels until 1981. Nevertheless, the question of surpluses, overcapacity and declining prices began to reemerge in the early 1980s, this time with Western Europe as surplus producer of several agricultural products. In response to high prices in the 1970s, farmers in most of the advanced countries expanded production significantly and began to compete aggressively with each other for international sales. By 1983, there was an increasing reluctance to support surplus agricultural stocks in the United States and in the Common Market countries. Declining world prices in the early 1980s had forced a heavy outlay of funds to maintain the minimum price levels of the Common Agricultural Policy in Europe, and the sharply rising expenditures for agricultural subsidies increased strains among the members of the European Economic Community. In time this will likely cause a restructuring of the CAP. The American approach was the Payment in Kind program introduced in 1983 and subscribed to by an unexpectedly large number of farmers. The effect of PIK, combined with unusually dry weather, was immediately to push up grain prices worldwide (easing the subsidy payment problem for the CAP) and to reduce U.S. grain stocks. This, however, was a short-term effect, and prices of agricultural products continued to slip through the mid-1980s.

Whereas the 1970s began optimistically and then experienced a sharp deterioration in the world food balance, the 1980s began with some anxiety but then moved in the direction of surpluses. The U.S. attempt to "fine tune" production through the PIK program in 1983 failed and is viewed as an overly expensive short-lived experiment. It is difficult to determine the effects of the acute financial problems facing some 10 to 15 percent of U.S. farmers in the mid-1980s. However, even if the land changes hands, it will not go out of production, and programs will probably be implemented to keep most of those affected in business.

Medium-term disturbances are by nature impossible to predict, but the events of 1972 demonstrate the inherent uncertainty of the balance between the supply and demand of world foodstuffs. Policy-makers should be alert to the error of projecting the status quo into the medium-term future, especially when this leads to significant run-down of stocks. It would be better to be prepared with contingent stocks sufficient to offset moderate supply disturbances. Stocks adequate to meet worst-case assumptions would be prohibitively expensive, but prudence calls for the maintenance of adequate grain stocks to avoid sharp peaks and valleys in supplies.

In addition to such contingent stocks, stockpiles of grain are also needed to meet short-term local shortfalls in specific areas, generally in Africa. Before 1950, such failures commonly resulted in famine, but the provision of basic foodstuffs on a concessionary basis has vitiated the effect of these disasters during recent decades. Because the short-fall is limited in area and is temporary, it can be fairly easily over-come out of existing stocks, providing ways of transferring them to areas of need are at hand and local authorities are cooperative.

Chronic Nutritional and Food Distribution Problems

Chronic undernutrition is more difficult to identify and remedy than famine. Estimates vary regarding the number of people in the world who are undernourished, but most analysts agree that between 25 and 30 percent of the populations in the Far East and Africa and 10 to 15 percent in the Near East and Latin America suffer from chronic undernourishment; in some countries it is of course much higher. Most of these nations are poor and have great difficulty providing remunerative employment or concessionary food to their poorest inhabitants.

The underlying reasons for chronic undernutrition are the basic, complex problems of underdevelopment, rapid population growth and low per capita income in the poorest parts of Asia, Africa and Latin America. This situation is aggravated by local uncertainties and losses in food production; crop sensitivity to variations in weather conditions; high loss of crops to pests in the field; lack of storage facilities; and poor transportation systems leading to spoilage and further losses. The solution to undernutrition is slow and difficult and centers on education, agricultural development, employment programs for the poorest groups, and reducing population growth.

Redistribution of foodstuffs from surplus to deficit countries on a concessionary basis could reduce the undernutrition problem in the short run. However, experience with concessional food programs in the 1950s and 1960s revealed some unanticipated problems. Such programs often reduce food prices in the recipient country and unfortunately depress local agricultural production. Local farmers facing lower prices for their products cannot afford as much fertilizer or pesticide and their yields decline accordingly.

Moreover, in the absence of successful development and population programs, concessionary food may only raise the population, resulting in continuing undernutrition, but with increased dependence on imported food. Concessionary food programs should be used only in combination with a broad and well implemented program for agricultural development; otherwise, it could become counterproductive over several years. The solution to chronic undernutrition is not concessionary food imports, but increased local output based on education of farmers; better land tenure conditions; improved agricultural inputs, especially seed; adequate storage and transportation facilities; and government policies that provide farmers with adequate incentives to produce.

Government Policies Affecting Agriculture and Food

Government policies are an important determinant of agricultural performance at the national level. In the developed countries, agricultural programs have aimed to support farm incomes at an adequate level and have created problems of surplus production, to be discussed below. In the developing countries, agricultural policies in the 1950s and early 1960s were aimed principally at extracting the surplus from agriculture and using it to accelerate industrial development. Sizable worldwide food stocks also worked to reduce any urgency regarding agricultural output, and in fact net imports of cereals by developing countries rose from an average of 4 million metric tons in 1948 to 25 million m.t. in 1964.[9] Developing countries deemphasized agriculture through two main policy choices. First, prices were set to keep food prices low for urban consumers and to extract value from the agricultural sector. This policy tended to depress growth in agricultural output, but shortfalls in supplies were typically made up through imports, often on a concessionary basis. The second policy choice was the relative lack of investment allocated to agriculture and the direction of resources toward industry, frequently heavy industry.

By the early 1960s, the policy of relegating agricultural development to a low priority was creating several serious problems. Growth in agricultural output was decelerating, while population growth remained high. Projections of existing trends showed a potential for food shortages developing in several countries. It was recognized that imports were not a permanent solution because of the foreign exchange costs implied by imports in excess of available concessionary flows. Within the agricultural sectors in the developing countries there were additional problems: increasing rural poverty; a rise in landless peasants; inadequate production of agricultural raw materials for the industrial sector; and the above-mentioned failure to keep up with domestic demand for low priced foodstuffs. During the 1970s, a growing awareness of the problems arising from deemphasizing agricultural development led to some reduction in policy-induced price distortions in food markets, greater emphasis on agricultural investment, and greater use of fertilizer, irrigation and higher yielding varieties of seed. The combination of fertilizer, irrigation and better seed has accounted for half the increase in grain yields in developing countries since 1950.[10] Fertilizer use rose tenfold to 46 million m.t. in 1981; irrigated land areas increased by about 2 percent a year; and agriculture received 15 to 20 percent of total public investment in the 1970s.[11] The fruit of these new agricultural policies is evident in the increased growth in food output that occurred in the 1970s — 3.5 percent after 1974 compared with 2.5 percent in the 1960s.

In the Soviet Union and several Eastern European countries, government agricultural policies have failed to maintain satisfactory growth in food output. In the 1960s, food output in this region grew at a strong 3.2 percent annually. This growth continued into the early 1970s, but then several poor harvests occurred. During the 1970s, food output in Eastern Europe and the Soviet Union increased only 1.7 percent per year. Poor weather was certainly a factor, but decades of directing investment toward heavy industry and military activities combined with inadequate incentives for producers have taken a large toll on agricultural production in these countries. The consequences of these poor policies are being felt worldwide because of Soviet absorption of grain on world markets. Compared with negligible imports in the 1960s, Eastern Europe and the Soviet Union have consumed an average of a quarter of available grain exports since 1973. Were it not for the failure of Soviet agriculture in the 1970s, the world food balance would be a matter of declining importance and the issue of surplus production even more urgent than it is.

Agricultural policy in the market-oriented developed countries has a history of trying to stabilize and to shore up farmers' incomes. In the European Economic Community, the CAP has resulted in the production of surplus amounts of many subsidized or supported crops and has continued to raise the issue of how surpluses can be disposed of without imposing too great a cost on public finances. In the United States, programs have tried to take a more market-oriented course, but have not succeeded. During periods of high crop prices, such a policy requires little or no public funding, but when prices fall, some action is required. Faced with such a need in the spring of 1983, as discussed briefly above, the U.S. government offered its Payment in Kind program. This guaranteed that farmers would receive a quantity of grain equal to their normal crop in return for holding their land out of production. Unfortunately, an unexpectedly large number of U.S. farmers agreed to participate, raising costs above anticipated levels.

Meanwhile, poor rainfall reduced yields on the acreage planted. The result was higher food prices, lower stocks and a costly, short-lived experiment in farm policy. The cost problems of the CAP and PIK emphasize the inherent difficulty of providing farmers with income security through price subsidies. In the mid-1980s, with agricultural prices even more depressed, market-oriented policies are unlikely to produce politically acceptable results. As a consequence, agricultural support programs continue at high levels without satisfying farmers, consumers or taxpayers.

While a successful formula for agricultural price support programs has not been found, the developed countries have been more successful in providing their agricultural producers with research and extension services. Publicly supported agricultural research totals between 1 and 2 percent of the value of agricultural output, while extension services are about 0.5 percent. In developing countries, overall support is lower, with the greater part of the spending going to extension services.[12] The rates of return are quite high for appropriate research into agricultural problems in developing countries, and levels of investment are inadequate. Over time, publicly supported research in the developing countries should be increased. Government policies make an enormous difference to agricultural results, and in the developing countries these policies have been moving in the right direction—fewer price distortions and more investment. The results achieved since 1974 prove the efficacy of the new policy approaches. In the Soviet Union, there is as yet no sign that agricultural policies are being significantly reworked, and as a consequence the Soviets are likely to continue to be a net drain on world food resources. In the developed countries, the steady growth in yields should continue and surpluses should rise, unless government policymakers induce a series of manmade supply shocks in the years ahead.

Long-term Supply Trends in Agriculture

During the last two centuries, agricultural supplies have grown faster than population and consequently per capita food consumption has improved. The improvement of diets has been greatest in Europe and North America, but since 1950 the evidence indicates some improvement in almost all of the developing world as well. This is a truly impressive achievement when one considers the growth in total world population from roughly 1 billion in 1800 to about 4.5 billion in 1980. A review of world crop production from 1950 to 1980 will bring the achievement of agricultural producers into proper focus (see Table 1–2). The figures indicate increases in the range of 150 percent for most crops, with sugar somewhat below this trend and soybeans well above. Soybeans have increased especially rapidly since 1970 for use as a

**TABLE 1-2. ANNUAL WORLD CROP PRODUCTION, 1950-80
(Million Metric Tons)**

Crop	1950	1960	1970	1980	Total Increase
Wheat	172	222	288	420	144%
Rice (paddy)	153	151	293	375	145
Corn ⎫				385 ⎫	
⎬	286	361	445	⎬	156
Other grains ⎭				347 ⎭	
Sugar	39	62	72	90	131
Soybeans	18	26	41	98	444
Cottonseed	12	21	21	30	150
Peanuts	10	14	17	25	150
Potatoes	242	247	257	285	18

Source: Food and Agriculture Organization, *Production Yearbook*, various years.

protein supplement for animal feed. Potatoes present an interesting case of a crop for which demand did not increase significantly, and as a consequence the area planted actually declined. The approximately 150 percent increase in production of the other crops listed between 1950 and 1980 is an annual compound growth rate of 3.1 percent. For the most part, these gains have been achieved through increased yields rather than by cultivating more land, and this is the great success of world agriculture since 1950. Focusing on cereal yields in the countries of the European Community since 1955, production in terms of tons per hectare has increased by an average annual growth rate of 2.6 percent, with wheat yields growing by 2.9 percent and corn yields by 3.7 percent.[13] The story is much the same in other regions of the world with cereal yields in the United States rising by an average annual rate of about 3 percent between 1960 and 1980 and annual yields in the developing world rising by about 2.5 percent over approximately the same period.

Total world food production increased at an annual rate of 2.4 percent through the 1960s and 1970s (see Table 1-3). The slowdown in the early 1970s was compensated by accelerated growth after 1974, so that food production grew at about the same rate in both decades.[14] This compares with annual growth in world population of just over 2.0 percent, indicating an improvement in per capita food consumption of about 0.4 percent per annum; however, these average figures hide significant differences among regions and within regions of the world. Since 1972 food production has accelerated, and the developing countries have done better than the developed countries. Only

TABLE 1-3. TOTAL FOOD PRODUCTION BY REGION, 1960-81 (1969-71 = 100)

	1960	1970	1974	1980	1981	Growth Rates		
						1960-70	1970-74	1974-81
World	80	100	110	125	129	2.3%	2.4%	2.3%
Developing Regions								
Africa	79	101	110	134	140	2.5	2.2	3.5
Far East	81	99	108	124	125	2.0	2.2	2.1
Near East	80	101	106	130	137	2.4	1.2	4.4
Latin America	78	101	114	133	136	2.6	3.1	2.6
	80	102	113	142	149	2.4	2.6	4.0
Developed Regions	81	99	110	119	121	2.0	2.6	1.4
United States	89	97	107	125	135	0.8	2.5	3.4
Western Europe	78	99	113	123	121	2.3	3.4	1.0
Oceania	77	99	110	122	131	2.5	2.6	2.5
Eastern Europe	74	101	113	116	115	3.2	2.8	0.3

Source: FAO, *Production Yearbook*, various years.

since 1975 has developing country food production moved noticeably above population trends, enabling additional improvement in diets. A breakdown of the developing countries by region, shown in Table 1-3, reveals the difficult agricultural situation in Africa and the great strides that have been made in the Far East and Latin America, particularly since 1974.

In Africa, food production has lagged behind population growth for more than 10 years. This is the one region of the world in which agriculture is not succeeding, and concessionary food aid will be needed for the foreseeable future to help alleviate large-scale starvation. In the Near East, food production growth has more than kept pace with population, but since 1978 this growth has fallen off, possibly due to the political and military turmoil prevalent in the region. Both Africa and the Near East are food deficit regions where an inadequate food balance is worsening. Both will require increasing agricultural development support in the 1980s if a serious deterioration in the food balance is to be avoided. Environmental problems, however, combined with widespread political instability and military insecurity make constructive and sustained action all but impossible. In the Far East and Latin America, the overall performance of the agricultural sector has been better. Growth in food output has increasingly outpaced population growth, and this trend has accelerated since 1974.

For the Far East developing countries, growth since 1974 moved up to 4.4 percent compared with 2.4 percent in the 1960s and 1.2 percent in the early 1970s. In Latin America, the output of crops and livestock grew by 4.0 percent since 1974 compared with 2.4 percent in the 1960s and 2.6 percent in the early 1970s. Progress within Latin America has been greatest in countries like Brazil and Argentina where strong incentives have encouraged agricultural producers to increase output prodigiously. In other countries in the 1970s, particularly in Mexico and many Central American countries, producers were unable to keep up with rising demand, and a growing food deficit and increasing imports resulted. In the future, higher prices for agricultural products are likely to reduce the growth of demand, resulting in greater exports and/or reduced imports. Nonetheless, the overall performance of agriculture in the Far East and Latin America shows the kind of progress that is possible when proper agricultural development programs are implemented. (These programs are discussed in more detail in Chapters 6 and 7.)

During the 1960s and 1970s, the expansion of arable and permanent cropland worldwide was negligible—a decline of 1.5 percent in the 1960s was offset by an increase of 2.8 percent in the 1970s. This increase is quite small compared with the total food output growth of 25 percent in each decade. The increase in land usage in the 1970s was attributable primarily to the developing countries, with 36 per-

TABLE 1-4. SOURCES OF CROP INCREASE,
DEVELOPING COUNTRIES, 1961-80
(Annual Rates of Growth)

Crop	Land	Yields	Output
Corn	1.0%	2.3%	3.3%
Rice	1.0	2.8	3.8
Wheat	0.4	2.4	2.8
Root	0.8	1.1	1.9
Sorghum and millet	0.7	0.9	1.6

Source: FAO, *Production Yearbook*, various years.

cent of the worldwide increase in Latin America and 30 percent in Africa. When broken down by crops, the greater contribution of yields compared with increased cultivated areas is clear (see Table 1-4).

Over this period, total cultivated land in the developing countries increased by 15 percent. There are programs under way for the continued settlement of new land in South America, principally in the Amazon region, and in parts of Indonesia. Such expansion is important, but higher returns are usually earned from improvements in existing agricultural practices and investing in existing farmland. Thus, new land cannot be expected to support required future increases, and improved yields per unit of land will more than ever be the dominant factor in world food production.

THE WORLD FOOD BALANCE IN THE YEARS AHEAD

The six determinants discussed above provide a framework for considering the likely prospects for the world food balance. The evidence is not all on one side; there are positive and negative trends at work, and there is some disagreement among experts as to the likely course of events up to the year 2000. Most analysts, however, come out with a guardedly optimistic prognosis.

The most important factor is the worldwide slowdown of population growth — a small downturn is discernible in most of the developing countries and there is a continued slowing of population growth in developed countries. Worldwide, the rate of population growth peaked in the 1960s, fell slightly in the 1970s and, according to World Bank analyses, is projected to fall even further in the 1980s and 1990s. These results are shown in Table 1-5.

The exceptions to the downturn in growth rates seem to be most African countries, Pakistan, Bangladesh, and a few high income oil-exporting countries. These oil exporters have small population bases

**TABLE 1-5. WORLDWIDE RATES OF
POPULATION GROWTH**

Decade	Growth Rate
1950s	1.90%
1960s	1.96
1970s	1.83
1980s	1.70
1990s	1.63

Sources: World Bank, IBRD; WDR, Population
Table, various years.

and can pay for their food imports, but in Africa, accelerating population growth in the face of declining agricultural output forebodes a growing food shortage for many years to come. Thus, merely to maintain existing dietary levels, Africa will need increasing inflows of food on a concessionary basis. Pakistan and Bangladesh could also be in need of large-scale food assistance, but Pakistan has made remarkable strides in raising its level of domestic food production and has more than offset population growth. Bangladesh has tremendous potential to raise domestic production by bringing rice yields up to the levels prevailing in several East Asian countries with less favorable climatic, soil and water control conditions and is beginning to show improved yields. In all the other regions of the globe, population growth is decelerating, and this trend should continue into the next century.

The second positive factor at work to improve the world food balance is the gain in output and yields experienced by most countries during the past 10 years. In all regions of the world—again except Africa—local food production per capita has increased by about 10 percent over the past decade. Among the Eastern European countries the results are mixed, with Hungary and Romania raising per capita output by 32 and 47 percent, respectively, while Russia increased per capita output by only 2 percent and Poland declined by 4 percent.[15] The market-oriented developed economies logged steady progress, raising their per capita food production in the 1970s by 12 percent. The United States achieved a 16 percent increase and Canada a 9 percent rise.[16]

Most of the larger Far Eastern and South American developing countries made striking gains in food output per capita during the 1970s; nine increased per capita food production by more than 15 percent, and nearly all the rest improved by 5 to 15 percent. This tremendous achievement represents a significant improvement in the nutritional intake of the citizens of these countries. See Table 1-6.

SEEDS

**TABLE 1-6. FOOD PRODUCTION PER CAPITA AND
CHANGE IN POPULATION GROWTH RATES,
SELECTED DEVELOPING COUNTRIES**

Country	Change in Food Production Per Capita, 1970-80	Change in Annual Population Growth Rates, 1960-70
Sri Lanka	48%	− 0.7%
Malaysia	39	− 0.4
Thailand	29	− 0.5
South Korea	26	− 0.9
Brazil	25	− 0.7
Philippines	22	− 0.3
Colombia	22	− 1.1
Indonesia	17	+ 0.2
Argentina	16	+ 0.2
China	16	− 0.8
Mexico	6	− 0.2
Pakistan	5	+ 0.2
India	3	− 0.2
Bangladesh	− 6	+ 0.1

Source: WDR, 1982 and 1983.

Many of these countries also increased cereal imports, indicating that nutritional levels have risen even faster than the gains achieved by domestic production. In the process, the governments have learned some important lessons regarding the need for price incentives for farmers, the need for channeling sufficient resources to agriculture, and the need for growth in agriculture to support industry and to conserve and earn foreign exchange. As attested to by the disappointing results experienced by other developing countries and by several centrally planned economies, these lessons are not learned easily, but it is to be hoped that, once learned, they will not be easily forgotten.

On the negative side, several factors emerge from a review of the determinants of the world food balance. The most important of these in terms of human misery is the deteriorating food balance in Africa. Environmental deterioration explains part of this and suggests that the declines will not be easily reversed. Most African countries are simultaneously accelerating population growth and producing less food per capita. A comparison of Table 1-7 for Sub-Saharan African countries with Table 1-6 for the larger Latin American and Far Eastern countries shows vividly the contrast in food balance trends among these regions of developing countries. The food balance in Sub-Saharan Africa is obviously worsening, and there is little chance of reversing

TABLE 1-7. FOOD PRODUCTION PER CAPITA AND CHANGE IN POPULATION GROWTH RATES, SELECTED AFRICAN COUNTRIES

Country	Change in Food Production Per Capita, 1970-80	Change in Annual Population Growth Rates, 1960-70
Ivory Coast	+ 10%	+ 1.2%
Central African Republic	+ 2	+ 0.4
Cameroon	+ 1	+ 0.4
Liberia	− 5	+ 0.3
Zambia	− 8	+ 0.5
Nigeria	− 9	0.0
Tanzania	− 9	+ 0.7
Zaire	− 13	+ 1.0
Kenya	− 15	+ 0.8
Ethiopia	− 15	− 0.4
Congo	− 18	+ 0.5
Senegal	− 24	+ 0.4

Source: WDR, 1983.

this trend in the 1980s. These countries will need increasing imports of food, and most will be unable to pay for what they need. This will almost certainly be the most difficult food problem facing the world in the next two decades.

Excluding Nigeria, which has oil revenues, and the Ivory Coast, which managed fairly well in the 1970s, approximately 30 countries with a combined population of 250 million will require increasing food imports on a concessional basis if famine and starvation are not to become endemic to Sub-Saharan Africa.

A second important factor is the increased demand that will result from income growth in the successfully developing countries as characterized in Table 1-1. Even when countries raise production faster than population increases, there is still a need for greater food imports because higher incomes enable people to improve their diets. Between 1961 and 1976, the 16 developing countries with the greatest increases in domestic food production more than doubled the volume of their food imports.[17] This demonstrates the power of the income effect on food demand in the middle ranges of development. To keep up with demand growth, it is necessary for a successfully developing country to raise food production per capita by 2.0 to 2.7 percent per year or by 22 to 30 percent per decade. Only the fastest growing countries listed in Table 1-6 are meeting this challenge. Moreover, because of changes in diets, demand for staple grains grows rapidly for use as fodder. Demand for traditional root crops and starchy foods tends to decline with rising income, but because of the dietary shift to meat, the derived demand for imported grain tends to rise rapidly for successful middle level developing countries.

The conclusion is that the world food balance will vary from about the same to modestly better through the 1980s, with specific regions showing serious shortfalls. There will be a growing demand for imports to improve diets in the middle income developing countries and the centrally planned economies. However, output should expand faster than population in all regions of the world, except Africa, and increases in demand based on higher incomes will probably not put pressure on existing production capacity. It is possible that structural changes within several potential exporters, such as Brazil and Argentina, could lead to increased supplies in world markets in coming years. This could lead to excessive surpluses and continued low prices.

THE ERA OF KNOWLEDGE IN AGRICULTURE

The manageability of the world food balance as discussed in the preceding section should not mask the fact that an era has ended—the era of cheap energy—and a new one has begun—the era of knowledge. Just as the closing of the land frontier in the nineteenth century had important consequences, so also will the end of the energy frontier. In this setting, improved varieties of seed become more important than ever. Their effect on agriculture over recent decades will be demonstrated and prospects for the future will be discussed.

Increased food production in the decades ahead will be even more dependent on knowledge—that is, genetic research, improved agricultural techniques and equipment, and better informed government policymakers implementing more appropriate policies. The foundations for the "knowledge era" in agriculture are already well established. Agribusiness firms have the capacity continually to improve their products through privately funded research. Public sector research is available to fill in where private activity is absent and to spur private companies continuously to upgrade the quality of their output. Enormous gains can be expected by dispersing appropriate agricultural techniques and knowledge throughout the developing countries. In some cases, this implies transfers among developing countries, and in other cases, the adaptation of techniques and knowledge currently used in the developed countries. Finally, government policymakers must learn from past mistakes and successes. In the developing countries and the centrally planned economies, agriculture needs adequate incentives and infrastructure to foster domestic production and keep imports at manageable levels. Developed countries must be able to support their agricultural producers in ways that do not impose excessive burdens on taxpayers and consumers, negate world market forces or lead to protectionist agricultural trade policies. All these are crucial challenges to be faced in the coming decades, and all require the application of better information to existing problems. World agriculture has truly moved into the knowledge era.

The End of the Era of Cheap Energy

For most of recorded agricultural history, production has been largely dependent on the quantity and quality of available land. This is not to say that agricultural techniques and implements were unchanged century after century, but that change came so slowly that it was usually imperceptible to all but the historian. Many analysts feel that it was the opening up of fertile frontiers in the Americas and in Oceania that invalidated the predictions of Malthus. The introduction of the plow to vast new tracts of rich land enabled world food production to grow during the eighteenth and nineteenth centuries as it never had before.

The United States is a classic case of agricultural expansion through the extension of cultivated land area. The goal of the farmer was to increase income, that is, output per man-hour, rather than land yields. The result was extensive agriculture with average farm sizes increasing and the farming process becoming more mechanized. The flat prairie lands of the Midwest were particularly responsive to the use of the steel plow, the reaper and the thresher. By 1925, the era of cheap land had ended, and the focus of U.S. agriculture shifted from increased use of land to increased output per acre. Australia and Argentina were among the last to bring their full endowment of cultivable land into production. However, the cheap land era can be dated as effectively ending with the first quarter of the twentieth century.

A new era in world agriculture, the era of cheap energy, had been anticipated by developments that began in 1910 and by 1940 had changed the nature of farming in the United States. This was the mechanization of agricultural production based on the availability of cheap fossil fuels. In 1925, about 4 percent of the 6.5 million farms in the United States had tractors, and 40 percent had some kind of motor vehicle.[18] By 1940, the horse and mule had been almost completely replaced both for working the land and for transporting crops to market. In the process, fossil fuels not only raised output per man-hour and yields per acre, but freed 60 million acres—15 percent of U.S. cropland—that once produced oats and barley to feed draft animals.[19] Today, in addition to fueling farm machinery and equipment, petroleum products provide the feedstock for nitrogen fertilizer, pesticides and many other chemicals used in farming. For 50 years, cheap energy made it possible to expand output while dramatically reducing the direct labor required in farming. But with the oil price increases of the 1970s, fossil fuels can no longer be counted on to provide cheap productivity gains. Just as land became a constraining factor at the end of the first quarter of the twentieth century, so energy has become an expensive input that must be conserved. Reflecting this change is the growing concern with the energy efficiency of particular crops and agricultural practices.

There is another dimension to the end of the cheap energy era for agriculture—the potential draw on agricultural production for fuels based on biomass. Thus far, only Brazil has made a serious effort to utilize agricultural output to reduce the use of fossil fuels. Because of its extensive endowment of sugar cane lands, availability of labor and low world sugar prices, Brazil has been reasonably successful in producing alcohol from sugar cane.

American attempts to use biomass focused on the distillation of ethanol from corn and other grains, which was then mixed with nine parts of gasoline to produce a fuel called gasohol. As with Brazilian sugar, converting grain to fuel becomes sensible only when the price for grain as food or livestock feed is very low and the price of petroleum is very high. However, the high use of energy in the fermentation and distillation processes makes grain based on ethanol an extremely problematic fuel. Cost estimates range from $1.65 to $1.80 per gallon of ethanol based on a corn price of $2.50 a bushel at the plant and energy prices that were prevailing in the early 1980s.[20] This is about double the production cost of a gallon of gasoline at a comparable point in the production chain (i.e., exclusive of distribution, retailing and sales tax expenses). The gasohol program was supported by a federal subsidy of $0.40 per gallon and a similar state subsidy in over 20 states. The subsidies, in the form of exemptions from sales taxes in most cases, were barely enough to make the process feasible. Higher subsidies are unlikely, and gasohol is therefore unlikely to be a significant factor in consuming grain and supplying liquid fuels in the United States.

Agricultural costs are inextricably tied to energy prices by input and output relationships that run both ways. Falling energy prices would reduce the cost of inputs into agriculture and terminate any diversion of grain into use as fuel. Rising energy prices would raise agricultural production costs and could increase demand for biomass as a fossil fuel substitute. The era has ended in which energy was cheap and could be used in greater amounts per unit of agricultural output, even as the price of agricultural output fell in real terms. In the new era, knowledge will have to be the base from which more output is derived from a given amount of energy, just as energy was used to obtain more output per acre in the last era. "We've come to the end of the liquid fuel frontier"[21] and must press ahead with the exploitation of the knowledge frontier.

The New Era of Genetic Research

Knowledge, as embodied in the form of better seed, more productive equipment and techniques, improved infrastructure, and appropriate public policies, has been an important part of agriculture for centuries. Until the twentieth century, its creation and application were sporadic, unorganized and dependent on particularly gifted or enlightened individuals. In the past 50 years, the creation and application of agricultural knowledge has become more continuous, organized and systematic; in its current form it is appropriate to call it research. Until 1975, agricultural knowledge had as its underlying reality the availability of cheap resources, notably fossil fuels, to increase both output per man-hour and per acre. Now, the goal is to increase output per man-hour, per acre *and* per unit of energy. Improved genetic knowledge embodied primarily in the seed used in agriculture is arguably the most potent form of knowledge available to agriculture. Moreover, this form of knowledge is likely to increase in importance as techniques of biogenetic engineering become better understood and more widely exploited in agricultural applications during the next 20 years. The analysis here will focus on seeds and their creation, production and distribution internationally. Little has been written on seeds for the nonspecialist, and it is therefore appropriate to take an in-depth look at what is probably destined to become the most important building block available to world agriculture in the knowledge era.

It is not possible to predict what genetic research will achieve in the coming decades, but a look back at the progress of the past two decades will demonstrate the potential that has been built up. U.S. historical data show very little improvement in yields in the century between 1840 and 1940. Whatever improvements were occurring in terms of seed selection and better agricultural techniques were offset by the introduction of marginal land. Considering corn, for example, acreage planted rose from 30.0 million in 1866 to 101.5 million in 1930.[22] In the early 1800s, the evidence indicates that crosses of varieties indigenous to different regions produced the yellow dent corn that dominates plantings today in North America and Europe. After this breakthrough there were no comparable advances until hybrid corn was widely introduced in the 1930s. Throughout the prehybrid period, farmer-breeders in the United States were selecting corn seed from the plants that seemed best suited to their needs. As a result, many varieties emerged with a range of maturities that fitted them to climates ranging over most of the United States. By 1880, the hybridization phenomenon was known, but the methods available for the production of hybrids were not suitable for commercial use.[23] It was not until the 1920s that a commercially practical means of producing hybrid corn seed was developed, and it was not until the mid-1930s that the

TABLE 1-8. ACREAGE AND RELATIVE YIELDS FOR U.S. CROPS, 1910-70
(Acres in Millions, Yields 1967 = 100)

	Corn		Wheat		All Crops	
	Acres	Yield	Acres	Yield	Acres	Yield
1910–15	101	36	50	51	330	56
1920–25	101	38	60	48	358	56
1930–35	103	32	54	47	348	51
1935	96	35	51	51	345	54
1936	93	24	49	49	323	45
1937	94	41	64	53	347	62
1938	92	41	69	51	349	59
1939	88	43	53	54	331	60
1940	86	42	53	59	341	62
1941	85	46	56	65	344	64
1942	87	52	50	75	348	70
1943	92	47	51	63	357	64
1944	94	48	60	68	362	68
1945	88	48	65	66	354	68
1950	82	55	62	64	345	69
1960	81	77	52	100	324	88
1970	66	91	44	120	297	102

Source: *Historical Statistics of the United States, Colonial Times to 1970* (Washington, D.C.: Bureau of the Census, 1975), Series 496, 501, 502, 503, 506, 507, pp. 510–512.

new seed was widely used. By 1938, over half of U.S. corn acreage was planted with hybrid varieties, and by the mid-1940s, virtually all U.S. production was based on hybrid seed. The effect was to end the century-long stagnation in U.S. corn yields and to lay the foundation for the knowledge era in world agriculture. Table 1–8 presents a look at this crucial period in U.S. agriculture.

Chronically poor weather depressed yields in the 1930–36 period. The poorest corn crop since 1901 occurred in 1930 and the 1934 and 1936 crops were 30 percent worse — the 1934 crop was the lowest output level in over 50 years and the lowest yield level (15.1 bushels per acre) in recorded U.S. history. After 1936, conditions improved somewhat, and yields began to exceed levels of the 1910s and 1920s. By early 1941, the effect of the introduction of hybrids was evident, and yields per acre have continued to increase.

A series of trials conducted in 1972–73 and 1978–80 attempted to identify the contribution of improved genetics to yield gains of U.S. corn production. The trials utilized 47 different hybrids introduced at various times and widely used by American farmers between 1935 and 1980. One nonhybrid variety extensively used in the late 1920s and early 1930s was also included. The hybrids were grown under controlled conditions with regard to soil, weather, moisture, fertilization, and appropriate techniques. Test results indicated that an estimated 80 percent of the total yield gains for U.S. corn since 1930 were due to improved seed, that is, genetic yield gains.[24] In addition to yields, several other desirable characteristics were improved by the introduction of the new hybrids; these characteristics included the ability of the plant to stand upright (for harvesting), to resist stress and disease and to tolerate certain types of insects. Other estimates place this contribution somewhat lower, at an average of one bushel per acre per year or 65 percent of the total yield gain between 1930 and 1980.[25] Tests of different designs conducted at different times and locations will generally show somewhat different results, but as an order of magnitude indicator, it would seem that over recent decades, 60 to 80 percent of U.S. yield gains in corn production are attributable to improved seed. Moreover, it is not clear that the genetic yield gains have been diminishing. Total yield gain rates fell from 4.3 percent per year in the 1950s to 2.5 percent annually in the 1970s, but most of this decline is attributable to the diminishing effects of added increments of fertilizer. During the late 1950s and the 1960s, the application of nitrogen fertilizer raised yields by almost two bushels per acre per year. By the 1970s, the fertilizer-based yield gains had fallen to an annual average of less than one-half bushel per acre.[26] This decline in fertilizer-based yields more than explains the decline in total yield gains. Therefore, it can be argued that genetic yield gains held steady or perhaps increased somewhat in the 1970s.

Wheat, the other great staple crop of the United States, also achieved impressive annual yield gains (averaging 2.5 percent per year) over the last 40 years. As indicated in Table 1–8, yields declined gradually in the period before 1935 as wheat cultivation moved onto progressively more arid land and farmers emphasized output per man-hour rather than yield per acre. Wheat yields rose more slowly than corn yields in the 1940s, but gains accelerated after 1950. This was partly due to reduced acreage, but primarily due to increased and improved inputs, including seed, fertilizer, pest control, crop rotations, tillage and moisture control practices, and machinery. Estimates for the contribution of genetics to total yield gains for wheat are lower than for corn, but they are consistent with lower levels of research expenditure on wheat in the United States.

In Europe, the story is much the same for both corn and wheat. Over the quarter century from 1955 to 1980, corn yields rose at an annual rate of 3.7 percent, while wheat yields grew by 2.8 percent annually.[27] These yield gains are similar to, but slightly higher than, annual rates of U.S. yield gains for wheat, probably indicating some spill-over of the benefits of U.S. research into hybrid corn and high yielding varieties of wheat in Europe. In addition to increasing yields, genetic research has substantially increased the zone in which European farmers can cultivate corn by introducing earlier (faster maturing) varieties. This has extended the intensive corn growing zone in Europe into north-ern France, Belgium and all of West Germany. Improved seed has en-abled farmers in these regions to substitute corn for less valuable fod-der and grain crops, an achievement that is not reflected in yield data, but is nonetheless an important advance for northern European farmers.

A complementary point pertains to the cost of the improved seed. In real terms, the costs have fallen, which means that the cost of seed has not kept up with inflation, i.e., the cost of everything else, despite the fact that current varieties are roughly twice as productive as seed sold 25 years ago. Wheat seed in Germany, for example, increased in price by 30 percent between 1950 and 1980, while the cost of living index rose 129 percent. Adjusted for the cost of living, this represents a decline of 43 percent in the real cost of seed.[28] If the gain in seed productivity is factored in, the decline is even more pronounced, 75 percent.[29] Cost histories for seed in the United States show broadly similar patterns. Seed prices generally rose more slowly than the U.S. cost of living index between 1950 and 1980, while quality gains were much greater than the average for all other products.

Wheat and rice are two well documented genetic successes that were widely applied in developing countries—the so-called Green Revolution—during the late 1960s and 1970s. The high yielding strains of wheat and rice were developed in the 1960s at two of the world's network of International Agricultural Research Centers (IARC); see dis-cussion in Chapter 3. These research centers took an indirect approach to increasing yields, and their success is indicative of the complexity of the tasks facing genetic researchers. Past efforts to develop wheat and rice varieties with greater grain yields had resulted in top-heavy plants that bent or fell over. Such lodging problems may affect roots, stalks or other parts of the plant. The structural problems of a plant—fixing itself in the ground, supporting leaf and grain mass or standing rigidly for mechanical harvesting—come under the category of lodg-ing. This can be particularly serious in mechanized harvesting. "In cereals, lodging at angles of recline of 45 degrees and 90 degrees at the ripening stage can lead to average yield losses of about 25 per-cent and 40 percent, respectively."[30] A second problem was that ef-forts to raise grain yields through increased use of fertilizer often resulted in larger grains and in lodging problems, or very sturdy stalks, but no significant increase in grain generation. A short strong stalk was needed that would not absorb increased applications of fertilizer, but could support a larger head of grain even when stressed by wind and rain. To address this problem, the research centers bred semidwarf strong-stemmed varieties of wheat and rice that could utilize additional moisture and fertilizer to produce higher grain yields and could resist lodging even with increased grain weight. The key was to find semidwarf varieties in nature through mass selection and then cross them with varieties known for a high output of grain. Some of the progeny will possess the desired genetic traits, and these can be multiplied for further improvement and ultimate distribution to farmers.

The resulting high yielding semidwarf varieties of wheat and rice, with appropriate combinations of irrigation, fertilizer and chemicals, have yields two or three times greater than with similar inputs to conventional varieties. By the end of the 1970s, these varieties had spread to more than half of the wheat acreage and a third of the rice acreage in the developing countries.[31] In India, for example, the new varieties of wheat were rapidly and widely accepted. Six years after introduction in 1966–67, high yielding wheat varieties were planted on more than half of the wheat acreage and accounted for almost 75 percent of the crop.[32] Indian wheat production doubled in this period to 23.4 million tons before the new varieties were infected with yellow rust disease. By the mid-1970s, strains of wheat had been developed to resist rust disease, and output resumed its growth. By the end of the decade, India was producing 33 million tons of wheat, and in less than 15 years had gone from being the world's second largest importer to being self-sufficient.[33] In the Punjab, yields went from 1.1 tons per hectare in 1960 to 2.8 tons per hectare in 1980. Total production of wheat rose from a trend level of 1.8 million tons in 1960 to a trend level of 7.8 million tons in 1980, reflecting both yield increases and additional land sown in wheat.[34] Incomes and savings increased, investment in wells and tractors rose and backward linkages to seed and fertilizer suppliers stimulated those industries. The overall result was truly remarkable, and the positive effects continue to ripple outward.

The new high yielding semidwarf varieties of rice were not accepted as rapidly as the wheat varieties because they required a reliable supply of water during the dry season. Where irrigation was available, the new rice strains increased yields and, because of their short maturity, enabled double or, in some cases, triple cropping. Varieties of rice are now in trial that would enable three crops of rice and a fourth crop of vegetables in one annual cycle. By focusing on the Punjab in India, it is possible to get an idea of what can be accomplished in the right circumstances. Rice yields rose from 1.5 tons per hectare in 1960 to 4.2 tons in the late 1970s, and production went up from about 300,000 tons in 1960 to 4.5 million tons in 1980.[35] Interestingly, rice production began to expand rapidly in the early 1970s when the high yielding wheat varieties became infested with yellow rust disease, and farmers make great efforts to shift land to high yielding rice wherever possible.

No comparable breakthrough has been made for corn varieties grown in the developing countries. Growing corn in the developing countries seems to require breeding programs for each of many crop areas, and the expense of such research, relative to the gain in total output, can be prohibitive. A second issue that appeared in the early varieties of high yielding wheat is disease resistance. Local varieties have evolved that are able to survive under specific local conditions, and new varieties may be especially susceptible to diseases and pests. The disease and insect problems become progressively more critical as cultivation moves toward the equator.

To resolve this problem, further breeding is required to combine the yield characteristics of the new varieties and the disease resistance of the traditional varieties. It must be realized that the area of production that can utilize the new varieties is a big factor in determining the level of resources that can be committed to genetic research aimed at solving a particular problem. As a result, creating and distributing improved corn varieties for use in the developing countries has been a slow process to date. New varieties have had the greatest impact in Argentina, Brazil, China, Kenya, and Zimbabwe, but no sweeping across-the-board advances have been made to parallel the "Green Revolution" in wheat and rice.

Research is making progress in creating improved varieties of other crops for use in the developing countries. Hybrid sorghum is now grown in China, India, Mexico, and other Latin American countries for human consumption. Progress on legumes has been slower, and care must be taken to encourage production of these nutritionally important crops despite the slow growth of yields. Tree crops, sugar cane and certain nonfood crops are important exports in many developing countries, and research on them has a long and generally successful history. Several export crops, including sugar, palm oil, rubber, and coconuts, have benefited from at least one significant genetic breakthrough in the past two centuries.[36] Over the 30-year period from 1950 to 1980, groundnuts (an important source of vegetable oil) enjoyed spectacular yield gains of 5.1 percent annually, outpacing corn, wheat and rice.[37] For many export crops, demand growth has been sluggish, and the benefits to yield gains have been manifested in lower prices to consumers rather than higher incomes for producers. Since genetic research is expensive, the allocation of resources between various crops must be weighed carefully to determine how benefits are likely to be divided.

High energy crops such as corn, wheat and rice have seen dramatic yield gains over the past several decades, while high protein crops such as legumes and soybeans have lagged somewhat. From a dietary point of view, breakthroughs in high protein crops would be most beneficial. Legumes are also very important in a crop rotational system because they fix natural nitrogen in the soil in a way that helps nitrogen-using crops and replenishes the soil. The high yielding varieties of wheat and rice have attracted developing country farmers away from vegetables with positive effects on total food production, but negative effects on dietary balance and growing reliance on expensive chemical nitrogen. A high yielding legume would help redress both these imbalances.

A final point concerns three important new developments that could begin to influence world food production in 10 to 20 years. These are genetic transformation, cell and tissue culture and new nitrogen-fixing techniques.[38] Genetic transformation was made feasible by the discovery of new ways to manipulate deoxyribonucleic acid (DNA), the chemical that transmits the genetic characteristics of plants and animals. The object of genetic transformation is to create entirely new plants or to endow existing varieties with desirable characteristics by laboratory manipulation of the genetic code in existing plant matter. When these experiments are followed by useful breeding and selection, significantly improved plants can thus be developed to meet the increasing need for agricultural production.

Cell and tissue culturing techniques enable the regeneration of full-scale plants from small pieces taken from existing plants. This has been done for hundreds of years with potatoes and apple trees, but the technique has become much more widely applicable. Genetic transformation usually requires assistance from certain cell and tissue culture techniques. These techniques also allow scientists to select new traits on a scale not possible with traditional methods or to reproduce individual plants, true to type, with speed and precision not previously possible. Clones developed in the latter way would have very high uniformity of yield, maturity and many other critical characteristics. The first impact is likely to be on tree crops, including oil palms, coffee, rubber, and many types of fruit. Because these plants are long lived and bear high value crops, there is high return to the effort expended in reproducing them.

Nitrogen fixation is achieved naturally by certain bacteria that are formed in nodules on the roots of leguminous plants of the pea family. The nodules provide the bacteria with a viable habitat, and the bacteria take up nitrogen from the atmosphere and fix it in chemicals that can be used directly by the plant hosting the bacteria. Chemical nitrogen is one of the most expensive of the major nutrients, and finding bacteria to fix nitrogen more efficiently or establish new kinds of symbiotic relationships with nonleguminous plants (such as wheat or corn) would be a major advance for world agriculture. Rhizobium bacteria are currently being propagated and selected for improved characteristics, such as higher rates of nitrogen fixation. These three important advances are on, or just over, the horizon. Any one of them could be as important for world food production as the introduction of hybrid corn or of high yielding varieties of wheat and rice. Only time will reveal their full importance.

It is possible to identify three different types of contributions that genetic research might be able to make in the next 10 years: (1) major breakthroughs based on biogenetic technologies, tissue culture, improved nitrogen fixing, or dramatically better seed varieties; (2) extensions of existing techniques and knowledge such as further yield advances in hybrid corn and high yielding wheat and rice varieties; and (3) the transfer of knowledge among regions of the world to utilize it more fully to improve crop varieties and agricultural systems. The chances are rather good that advances of the second and third type will continue at roughly their current level. However, it is much harder to predict breakthroughs. Biogenetic technologies are still in the basic research phase and are not likely to make an impact on the world food balance until the 1990s at the earliest. This same time frame also applies to nitrogen-fixing techniques. Breakthroughs in tissue culture could influence trends in the late 1980s, but it is impossible to predict exactly how. It is known that research in agriculture has had very high rates of return historically, well in excess of the 10 to 15 percent real (after inflation) rates of return that business firms look for in their investment programs. The evidence strongly suggests that genetic research is being inadequately supported and that additional resources committed to this use would be repaid manyfold. Moreover, the products of genetic research are commonly underutilized through poor policy decisions and shortsightedness. As world agriculture moves into the knowledge era, both these problems must be confronted and overcome if improvements are to continue in the world food balance.

FOOTNOTES, CHAPTER 1

1. See, for example, *A Distant Mirror* by Barbara Tuchman (New York: Ballantine Books, 1978), p. 119.
2. Phyllis Deane and W.A. Cole, *British Economic Growth: 1688–1959* (Cambridge, 1967), p. 8.
3. French and German data from J.H. Clapham, *Economic Development of France and Germany* (Cambridge, 1966), pp. 5 and 278.
4. In France, for example, population rose from 36.2 million in 1871 to only 39.5 million in 1910. (Both figures exclude Alsace and Lorraine.)
5. For declines in infant and child mortality in low income countries, see World Bank, *World Development Report* (WDR), 1984, Table 23.
6. Ibid., Table 20.

7. See, for example, *Feeding the 5000 Million* (Amsterdam: ASSINSEL, 1981), p. 3.

8. Timothy Josling, *Problems and Prospects for U.S. Agriculture in World Markets* (Washington, D.C.: National Planning Association, Committee on Changing International Realities, 1981), p. 16.

9. J. Sneep and A.J.T. Hendricksen, *Plant Breeding Perspectives* (Wageningen, Netherlands: Center for Agricultural Publishing and Documentation, 1979), p. 4.

10. WDR, 1982, p. 67.

11. Josling, pp. 18 and 19.

12. WDR, 1982, p. 69.

13. *Feeding the 5000 Million*, p. 14.

14. Josling, p. 18.

15. WDR, 1983, p. 159.

16. Ibid.

17. John Mellor, *Food Outlook for the Developing Countries* (Washington, D.C.: IFPRI, 1982), U.S.E.A. presentation, p. 6.

18. G. Barger and H.H. Landsberg, *American Agriculture: 1899–1939* (New York: NBER, 1941), p. 204.

19. Harold F. Breimyer, *Preparing for the Contingency of Intense Pressure on U.S. Food Producing Resources* (Washington, D.C.: National Planning Association, 1982), p. 2.

20. Ibid., p. 13.

21. Vernon Ruttan, as quoted in Breimyer, p. 3.

22. *Historical Statistics of the United States, Colonial Times to 1970* (Washington, D.C.: Bureau of the Census, 1975), Series 502, pp. 511 and 512.

23. Sneep, p. 190.

24. "Genetic Contribution to Yield Gains of U.S. Hybrid Maize, 1930–1980," Donald N. Duvick, Pioneer Hi-Bred International, Inc., Johnston, Iowa, 1981.

25. Report of the 1982 Plant Breeding Research Forum (Pioneer Hi-Bred, 1983), "Agricultural Research as an Investment: Past Experience and Future Opportunities," Vernon Ruttan and W. Burt Sundquist, p. 79.

26. Ibid., p. 86.

27. *Feeding the 5000 Million*, p. 14.

28. Ibid., p. 4.

29. Implicit in this calculation is the assumption that other goods have not improved in quality since 1950; therefore, 50 percent probably states the true decline more accurately.

30. Sneep, p. 242.

31. WDR, 1982, p. 69.

32. Sneep, p. 21.

33. WDR, 1982, p. 69.

34. Ibid., p. 70.

35. Ibid.

36. Ibid., p. 71.

37. Sneep, p. 138.

38. This paragraph is a summary of material presented in WDR, 1982, Box 6.4, "New Frontiers in Agricultural Science," p. 64.

An Introduction
to Seed Production

2

The background information on the production and distribution of seed presented in this chapter provides a convenient way to raise several key issues facing the worldwide seed industry. The first section describes the preservation and utilization of the diverse genetic resources that occur in nature. The second section introduces essential information on plant breeding. The third section reviews legal and economic issues that seed production has in common with other knowledge-intensive endeavors. Some technical and legal information on the seed industry is a necessary preliminary to understanding the trade and other issues to be discussed in subsequent chapters. This chapter outlines such information for nonspecialists; those who are knowledgeable in these technical aspects of the seed industry may wish to go directly to Chapter 3.

GENETIC RESOURCES

For the greater part of their existence on earth, humans were hunters rather than cultivators. Agriculture is a comparatively recent development dating back only about 10,000 years when humans began cultivating crops for their value as food sources. Until that time, plant varieties evolved through natural selection processes, that is, through natural mutations and recombinations. This evolutionary process created a multitude of plant species, well adapted to their local environments, that make up the pool of primary genetic material available for cultivation.

Plant Selection and Genetic Variability

Once humans began domesticating and cultivating certain species, plant selection based on traits desired by humans was initiated. The earliest plant breeding practices probably consisted of saving seed from the best individual plants at harvest in the hope that next year's crop would have the desirable traits of the parent plants. Almost certainly agriculture began and evolved independently in separate locations, notably in the Near East and in Central America.[1] The first plants domesticated were cereals—specifically, wheat and barley in the Near

East and corn in Central America—and leguminous vegetables. Tree crops such as olives and dates were also important in the Near East, while tomatoes, cassava and papaya were important in Central America.

Over the millenniums, plant selection was based on the parallel progression of agricultural and natural processes. This generated a great number of varieties well adapted to specific localities and agricultural practices. As a result, there exists an enormous visible variation within the same species of many plants, implying an even greater genetic diversity.

The first systematic efforts to survey the origins and diversity of crop varieties worldwide were undertaken by N.I. Vavilov in the 1920s. He found certain zones containing a wide range of environmental conditions where the most important crops originated and were first domesticated. Further research has more accurately identified these areas, shown in Figure 2-1, although they are not substantially different from the zones found by Vavilov.

The ongoing increase in genetic diversity of plants proceeded until about 25 years ago. Many agricultural units were still self-sufficient and, particularly in the major zones of origin, geographic barriers acted to separate and isolate seed populations. During the past two or three decades, however, forces have been at work to reverse the expansion of genetic diversity, such as hydroelectric and irrigation projects, industrial development, roads, residential sites, and, most important, the spread of commercial agriculture. The last has led to the progressive standardization and normalization of cultivated varieties. Mechanization of agriculture requires that crops have uniform characteristics, and the tendency is toward universal plant models that are best suited for mechanical cultivation.

It is no longer possible to take for granted the existence of an ample spectrum of genetic resources, i.e., crop germplasm (seed for use in breeding programs), in the zones of origin of the various crops. Surveys undertaken in the 1950s and again in the 1970s seem to indicate the disappearance of many wild and weedy varieties of crop species from their natural habitats, especially wheat types indigenous to the Near East.[2] One of the negative results of the "Green Revolution" has been the rapid replacement of numerous native varieties of wheat and rice with high yielding semidwarfs. The gain in terms of yields is indisputable, but the loss in terms of unique combinations of germplasm is considerable.

Unimproved wild and weedy varieties are irreplaceable because they embody combinations of genes that have been well adapted to particular environments through long periods of natural and agricultural selection. Such combinations of genes cannot be readily reconstructed, even by the most modern breeding programs. The gene com-

FIGURE 2-1. ORIGINS OF THE MOST IMPORTANT CROPS

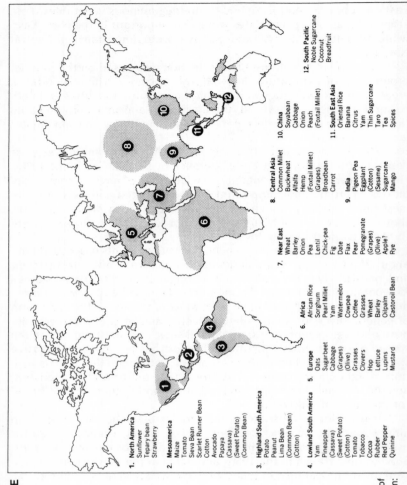

1. North America
Sunflower
Tepary bean
Strawberry

2. Mesoamerica
Maize
Tomato
Sieva Bean
Scarlet Runner Bean
Cotton
Avocado
Papaya
(Cassava)
(Sweet Potato)
(Common Bean)

3. Highland South America
Potato
Peanut
Lima Bean
(Common Bean)
(Cotton)

4. Lowland South America
Yam
Pineapple
(Cassava)
(Sweet Potato)
(Cotton)
Tomato
Tobacco
Cocoa
Rubber
Red Pepper
Quinine

5. Europe
Oats
Sugarbeet
Cabbage
(Grapes)
(Olive)
Grasses
Clovers
Hop
Lettuce
Lupins
Mustard

6. Africa
African Rice
Sorghum
Pearl Millet
Yam
Watermelon
Cowpea
Coffee
Grasses
Wheat
Barley
Oilpalm
Castoroil Bean

7. Near East
Wheat
Barley
Onion
Pea
Lentil
Chick-pea
Fig
Date
Flax
Pear
Pomegranate
(Grapes)
(Olive)
Apple?
Rye

8. Central Asia
Common Millet
Buckwheat
Alfalfa
Hemp
(Foxtail Millet)
(Grapes)
Broadbean
Carrot

9. India
Pigeon Pea
Eggplant
(Cotton)
(Sesame)
Sugarcane
Mango

10. China
Soyabean
Cabbage
Onion
Peach
(Foxtail Millet)

11. South East Asia
Oriental Rice
Banana
Citrus
Yam
Thin Sugarcane
Taro
Tea
Spices

12. South Pacific
Noble Sugarcane
Coconut
Breadfruit

Source: Reproduced by permission of *Feeding the 5000 Million* (Amsterdam: FIS/ASSINSEL, 1981), p. 6.

binations that exist in nature are truly the building blocks of the seed industry, and the loss of genetic resources is a potential setback for breeding capabilities that could have enormous consequences for future gains in agricultural yields. The greater the genetic variability existing in a population, the greater the range of action open either to natural selection or to breeders.

Narrowing of the Genetic Base and Crop Vulnerability

The other side of this problem is the progressive narrowing of the genetic base of crops worldwide and the consequent vulnerability of these crops to specific pests and diseases.[3] The history of agriculture is marked by recurring examples of an excessively narrow genetic base leading to crop vulnerability and significant losses. In some cases, the crop losses have been short term and limited in their overall effect; in other cases, the crop failures have resulted in catastrophic losses of life. Two examples of short-term crop failures without significant loss of life are Cuba's loss of more than $15 million worth of sugar cane in 1979–80 and the loss of 50 percent of the corn crop in the southern United States in 1970. In both situations, the affected crops were based on varieties that proved to be very susceptible to specific diseases; and in both, the problem was solved within two years by improving the genetic base of the affected crop so as to introduce greater resistance. These losses could have been avoided if the crops had been based initially on a greater diversity of genetic material.

Probably the greatest catastrophe to have resulted from an excessively narrow gene base was the Irish potato blight and resulting famines that occurred between 1840 and 1850. Potatoes had been brought from their zone of origin in the Andean region of South America in the sixteenth century and in less than 100 years had become the basic staple in Irish diets. The introduction of the new crop had triggered an enormous growth in population because yields in terms of digestible energy were on the order of three times as great as the crops replaced. Also, the Irish economy—under England's control—was unable to adjust. Markets and trade patterns were not geared to bring food into Ireland, and the political will to solve the problem was also lacking. The entire Irish potato harvest was based on very uniform genetic material and was therefore extremely vulnerable to a specific disease. Conversely, if the potato crop had been based on a wide genetic base, only a small part of it would have been vulnerable to any specific disease. Ultimately, the problem was solved by locating primitive varieties of potatoes in South America with resistance to blight and introducing those varieties into cultivation in Ireland and the rest of Europe. Within a decade, the problem of blight was under control, but several consecutive crop failures had resulted in the death or emigration of almost 30 percent of the Irish population between 1845 and 1850.[4] This tragedy demonstrates dramatically the importance of having a diverse supply of basic genetic resources where breeders can seek out necessary characteristics.

There have been innumerable other examples of crop failures due to genetic vulnerability, and such problems will most assuredly occur again. Several important crops in the United States and Canada may be vulnerable because of their narrow genetic base. Table 2–1 shows the percent of total crop planted to the six leading varieties in 1970 and 1980. To have only six varieties accounting for 50 percent or more of an important crop implies vulnerability and emphasizes the need for a broad reserve of backup varieties that can be searched for resistance to particular diseases and other useful characteristics. Fortunately, in all of these important crops, vulnerability seems to be falling over time, as the 1980 data show.

The problem of crop vulnerability due to a narrow genetic base is worldwide. The spread of high yielding varieties of wheat and rice in Asia, Africa, and Latin America is having the effect of narrowing the gene base of the most important crops in those regions. The first varieties of high yielding wheat introduced into India in the 1960s, for example, showed vulnerability to rust and had to be crossed with resistant local varieties in the early 1970s to ensure continued performance. In Italy, over 50 percent of corn production in the early 1980s was based on a single hybrid variety. At that time, the Italian corn crop was only potentially vulnerable, and fortunately a broadening of its genetic base seems to be under way. The potato crop in Europe continues to be based on a narrow genetic foundation due to the small number of varieties originally brought from South America and the loss of certain lines to disease. Potato blight is under control, but remains a concern for breeders as new types of the fungal disease evolve.

**TABLE 2-1. PERCENTAGE OF CROP PLANTED
TO SIX MOST POPULAR CULTIVARS**

Crop	1970	1980
Cotton	68	38
Soybeans	56	42
Wheat	41	38
Corn	71	45

Source: Reprinted by permission of *Economic Botany* (Vol. 38, No. 2), p. 163, D.N. Duvick (Bronx, N.Y.: The New York Botanical Garden).

It is in the tropical and subtropical regions where a single crop (such as rice) is uniform and grown continuously that the need for genetic diversity is most acute; pests and diseases can rapidly evolve into major problems and cause food shortages on a local or regional scale.

Cataloguing and Preserving the Genetic Pool

Although attempts to catalogue, collect and conserve seeds date back to the early 1800s, systematic and organized efforts to maintain genetic resources are quite recent. The potential gene reservoir is still very large for most crops, including varieties under cultivation, varieties used or stored in breeding programs, and diverse varieties existing in nature in the zones of origin.

The genetic resources that are potentially valuable can be categorized as follows:

Cultivated varieties

Commercial varieties, generally the result of breeding programs, show maximum response to intensive cultivation and complementary inputs of fertilizer, pesticides and so forth. In some cases, they may be subject to genetic vulnerability due to their narrow genetic base. They produce uniform plants and high yields under appropriate conditions.

Local traditional varieties, which have evolved through agricultural and natural selection processes, demonstrate wide diversity of characteristics, are well adapted to local conditions and do not produce uniformly high yields.

Obsolete varieties may have originated naturally or through breeding programs, but have been replaced as crops by newer varieties with one or more superior characteristics.

Special-purpose varieties have been selected for their ability to satisfy small, special-purpose demands, e.g., potatoes for starch or corn for human consumption.

Varieties in breeding programs

Breeding stocks are the elite genetic material used by breeders to produce and improve the leading commercial varieties.

By-product varieties are obtained as a result of breeding activities and genetic research, and include mutants, genetic testors and other by-product seeds.

Created intergeneric hybrids are combinations of pure lines of related species that result in varieties having some of the characteristics of both species.

Varieties in zones of origin

Wild species utilized but not cultivated by humans include trees and other types that grow wild and are used for their fruit, lumber, resin, or other products.

Primitive varieties of currently important crops may at some point be needed as genetic sources of disease resistance, dwarfism or other desirable characteristic.

Natural hybrids between species are used as crops and their wild relatives.

Wild relatives of current crops may have potential use in producing hybrids with useful characteristics.

Related species may provide useful genetic resources as breeding technology improves in the future.

Potentially useful species not now utilized as crops but which could be important in the future include fast growing trees, oil producing plants or species having use as raw materials for medicines.

In considering these categories, there is an ample store of commercial and special-purpose varieties and the genetic combinations used in breeding programs. Obsolete varieties are also adequately conserved by public agricultural institutions and seed companies in the developed countries.

In the developing countries, the obsolete varieties are local traditional varieties, and this category of genetic resources is deservedly receiving maximum attention in terms of collecting and conserving the diverse varieties used in traditional agriculture. Less attention has been directed to categorizing, collecting and storing the wild and primitive varieties still existent in the zones of origin. This will increasingly be the focus of conservation efforts in the years ahead.

The need for systematic and intensive international programs to collect and preserve the world's genetic resources is unquestioned; it is now a race against time to preserve many threatened plant species and varieties. As noted above, during the past two decades, the spread of commercial agriculture, mechanization and high yielding varieties in the important zones of origin has lent a new urgency to genetic conservation. On the one hand, the spread of the new high yielding varieties has pushed out the diverse traditional varieties that had evolved over the centuries to withstand local diseases, pests and climatic variability. On the other hand, the spread of the more uniform high yielding varieties, with their narrow genetic base has made crops more vulnerable to significant losses from a single disease or pest. The key to minimizing such losses is ready access to a broad pool of diverse varieties so that a resistant strain can be crossed with the high yielding variety to create a new variety that is both disease resistant and capable of producing high yields.

In response to the need for concerted action in maintaining genetic resources, the International Board for Plant Genetic Resources was established in 1974. The responsibilities of the IBPGR cover all activities concerned with preserving genetic resources, including " . . . identifying needs in exploration and collection, promoting and assisting field collection, improving and safeguarding storage of collected samples, training genetic research workers, promoting information exchange, improving data storage and retrieval systems . . . , advising on financial needs . . . ,"[5] rejuvenating and multiplying stored seed, and the evaluation of genetic material for its potential as breeding stock. Ultimately, the IBPGR should be backed up by a system of research centers at the international, regional and national levels for each of the important crops. The centers should work to gather, conserve, improve, and distribute germplasm to private breeders and to national research centers that can multiply and distribute the seed to the local agricultural community. A system of this type is in place or being developed for many important crops, including wheat, corn, rice, barley, sorghum, millet, potatoes, yams, and some legumes,[6] but there is still work to be done for other crops. The establishment of the IBPGR has accelerated progress in preserving genetic resources worldwide and, hopefully, support will build for greater efforts as the importance of conserving genetic resources becomes more widely understood.

Prior to the establishment of the IBPGR, national efforts had succeeded in cataloguing and collecting the principal commercial and improved varieties for the leading crops in the major producer countries. As international institutes were established for key crops in the 1960s and specimens from national collections were collated, it was found that very few minor, special-purpose, mutant, or wild varieties were represented. In the case of rice, corn and wheat, the international institutes instigated intensive field collections in search of traditional and special-purpose varieties. As a result, the holdings of the International Rice Research Institute (IRRI) rose from 14,000 varieties in 1971 to more than 53,000 varieties in 1978.[7] With regard to corn, some 18,000 varieties had been collected by 1955 through a coordinated Western Hemisphere program begun in the early 1950s.[8] Since the establishment of the IBPGR, efforts have concentrated on collating, rejuvenating and supplementing the varieties collected in the 1950s. This required additional exploration in South America, which focused on minor, special-purpose, wild, and mutant varieties. In addition, efforts were begun to acquire samples of the many unique varieties of corn found in southwest China and the eastern Himalayas that were not well represented in existing collections. By 1983, these and other undertakings had increased the number of corn varieties held by the Centro Internacional de Maiz y Trigo (CIMMYT), the international corn and wheat center, located in Mexico, to 14,000.[9]

The work of collecting and preserving the earth's endowment of genetic resources is painstaking, detailed and undramatic, but the importance of these efforts cannot be overstated. It is critical that existing germplasm be preserved and safeguarded if breeders are to be successful in developing the new varieties needed to feed the world's population in the years ahead. Two underlying principles for policymakers in dealing with seed and the seed industry should be:

- encourage the preservation and use of genetic resources worldwide; and
- encourage the broadening of the genetic base of important crops to reduce widespread vulnerability to specific diseases or pests.

In operative terms, these principles mean that governments should support and encourage responsible genetic research, both public and private, and should welcome the introduction of improved genetic material into the pool of varieties available to agricultural producers in their countries. Equipment and materials used in breeding programs should be exempt from import prohibition, as these are an integral part of the process that creates improved varieties and introduces them into large-scale use by farmers. If the creators and suppliers of high yield, disease resistant agricultural varieties are to succeed, policymakers must give their work and their needs an appropriately high priority.

BASIC PLANT BREEDING

This section will discuss important plant breeding facts and issues. To understand the production and trade of seed, it is essential to have some basic information regarding the types of seed produced, the range of characteristics that breeders are trying to embody in improved varieties, some of the methods utilized by plant breeders, and the time and cost involved in creating and marketing new plant varieties.

Propagation Properties of Plants

In terms of classifying plants from the breeder's perspective, it is important to note the following propagation properties of plants: vegetative propagation, where a cutting can be taken from an existing plant to reproduce a new plant with the identical genetic makeup; dioecious plants, where male flowers containing the stamens and yielding pollen and the female flowers containing the pistils are found on different plants; monoecious plants, where each plant typically possesses both male and female flowers and self-fertilization is possible. Within the latter group, plants can be further subdivided into autogamous crops and allogamous crops. Autogamous crops, in a natural setting, are typically self-pollinating and include the important crops of wheat, rice, barley, and soybeans. Allogamous crops reproduce principally through cross-fertilization in natural settings; this category includes corn, rye, sugar beets, and sunflowers. It is possible to control the fertilization of allogamous crops, but this requires time, money, technical expertise, and a systematic approach, i.e., a breeding program. And, as is discussed below, the greatest gains in crop improvement require that the plants be susceptible to both inbreeding, through self-fertilization if possible, and to controlled crossbreeding.

Historical Crop Improvement

Historically, the earliest form of crop improvement was mass selection where grains from the most vigorous and highest yielding plants were retained as seed and planted en masse the following year. Success was based on accurate selection for the desired types and a good mix of the available genes. With this approach, improvement depended on raising the average level of the gene pool, rather than on the fixation and maintenance of a specific superior variety. The advantages of mass selection are that the simplest and lowest cost methods are used, the generation interval is minimal (thus, a small yearly gain is equivalent to a much larger gain from a breeding cycle lasting three to six years), and there is good retention of the genetically superior material in the population.[10] The most significant disadvantages are that selection is based on appearance, not true genetic characteristics, selection is limited largely to whatever genes happen already to be at the location, and a farmer-breeder of cross-fertilized crops may be affected by whatever varieties neighbors are using. Mass selection is therefore the most basic approach to improving crops through genetic selection and provides a low cost benchmark against which more sophisticated breeding programs can be compared. It must be emphasized that such comparisons are much more than analytical exercises. In determining what to use for next year's seed, the cultivators of most crops clearly have the choice of retaining the best grain from the current year's harvest. Therefore, the products sold by seed companies are constantly being evaluated in terms of their cost effectiveness compared with the alternative of mass selection. The next level of evolution in breeding was to select and test the progeny for specific characteristics and then to repeat the process to get continuing improvement. For example, the ear-to-row selection method for corn was developed to evaluate the performance of the seed tested. A portion of the kernels from specific ears are planted in a separate row and the productiveness of each ear is established by the productiveness of its row. Seed from the ears showing the best test results is then multiplied for widespread use. Over the years, the ear-to-row method was improved by using check rows and replicating earlier experiments. Inbreeding was reduced by removing the male flower (detasseling) from the plants in the rows being tested. Experiments comparing ear-to-row selection methods with mass selection showed improvement in several characteristics, including oil content, ear length and maturity, but the superiority of ear-to-row selection in improving yields is not as clear cut.[11] However, the introduction of progeny testing, of which ear-to-row selection is an example, was a definite advancement in plant breeding.

Methods Utilized by Plant Breeders

Vegetative propagation is characteristic of crops such as potatoes, cassava and sugar cane where cuttings properly taken from a parent are capable of growing into a new plant with the identical genetic makeup. This is reproduction by nonsexual processes and is the natural equivalent of the new tissue culture techniques described in Chapter 1. Improvements in vegetatively propagated species are currently achieved through a careful selection of the parent plants, propagation of a large population of offspring, and picking the best of these in terms of the desired characteristics (e.g., yield, disease resistance). The approach is an up-to-date version of the ear-to-row technique. This process is repeated with increasing selectivity until the breeding program is judged to have accomplished its ends, and propagation for widespread distribution can begin.

Vegetatively propagated crops run the risk of having diseases transmitted from generation to generation to a much greater extent than crops propagated through seeds. Viral diseases in particular seem to be transmitted to the offspring via the cutting taken from the parent plant, whereas seeds protect against this occurrence. The production of virus-free potatoes, for example, for use as parent plants is very important and is possible in a carefully managed breeding program. It is also possible to cross-fertilize vegetatively propagated crops through techniques that encourage the growth of female and male flowers. Pollen is then taken from the male parent and applied to the stigmas of the female parent, which is then labeled. The resulting seed can be stored for an extended period or sown in the next planting season. This process must be repeated for many plants to ensure a large enough sample of seed to provide a high probability of ultimately producing varieties with the desired characteristics. In subsequent years, the seed is grown and the best potato from each plant is saved; the best potatoes from each plant are then replanted. Field tests proceed for five more years with increasing selectivity for both the desired breeding characteristics and for virus-free plants.

Breeding programs for autogamous, or self-pollinating, crops are based on a system involving three stages: creation of genetic diversity through the introduction of new varieties with a desired characteristic (e.g., dwarfism in wheat or rice); repeated selection and reproduction of lines possessing desirable characteristics; purification and maintenance of the improved varieties. Genetic variation is achieved by crossing two different varieties with desirable characteristics; for example, several wheat varieties with good local adaptability would be crossed with new varieties possessing the desired characteristic. After the crosses are made and the seed harvested, the first generation plants are grown, allowed to self-pollinate, and their seed used to produce the second generation of plants. At this point, selection begins and proceeds through four or five generations until the breeding material has been properly evaluated, and the strains with good genetic potential identified and isolated. By the fourth or fifth generation, desirable varieties have also been tested for disease resistance, and the purification process begins in earnest. Typically, it requires a total of 10 to 12 generations to prepare a new variety of wheat for official trials and testing. While the object of the breeding program is to introduce a desirable new characteristic, such as dwarfism, it is important that yield also be a criterion for selection at an early stage in the process. Another important selection criterion should be vigor in dense populations rather than performance solely in spaced plants.[12]

While necessarily brief, this description illustrates the system that has produced improved varieties of high yielding dwarf wheat and rice. The same breeding system would also be used to create varieties with disease resistance. The foundation of the whole process is the availability of varieties in nature that possess the required characteristic, hence the fundamental dependency of modern breeding programs on genetic diversity in nature.

After the genetic diversity has been introduced, the breeding program utilizes an updated version of the selection process that farmer-breeders have used for many years. The selection is much more intensive and systematized than mass selection, but the principles and goals are the same. The final steps—testing for disease resistance, purification of the seed line and maintenance of the variety—are aspects of a breeding program that farmer-breeders cannot reliably duplicate. These final steps provide an important degree of security to the cultivator who purchases seed and ultimately to the consumer who is dependent on a reliable and secure food production system. The basic breeding procedure used to improve self-pollinating crops—i.e., introduction of genetic diversity, selection, purification, and maintenance—continues to be a productive tool at the service of world agriculture. For certain crops such as corn, both cross-fertilization and self-fertilization can be controlled, and this led to the development of hybrid breeding techniques. During the nineteenth century, both accidental and experimental crosses between diverse corn populations produced startlingly positive results. In 1846, the Reid family emigrated from Ohio west to Illinois, bringing seed in a covered wagon. Their first harvest was inadequate to supply their corn seed needs for the next year, so they bought a local Illinois variety to mix with their Ohio variety. The cross-pollination of these diverse strains resulted in a hybrid stock of corn seed that was especially favorable.[13] Over the years, the Reids put the resulting seed through a rigorous mass selection process and developed a strain since used as the basis for many successful hybrids. Experiments conducted in the 1870s verified the powerful results of crossing diverse varieties of corn, while experiments in subsequent years established the harmful effects of inbreeding. But further tests showed that crosses of the right inbred lines would produce offspring markedly superior to either of the parent lines or to open-pollinated varieties—and this is the key to hybridization.

The phenomenon of hybrid vigor has come to be called heterosis, and while the basis for this effect is still not fully understood, its power is undeniable. With regard to corn, "inbreds typically are less than 30 percent as productive as their hybrid progeny."[14] This implies a typical heterosis effect of 3 to 1 over the yield levels of the parent seed for corn! The superiority of hybrids compared with the best nonhybrid varieties (as opposed to their less vigorous inbred parents) is in the range of 15 to 25 percent, a considerable gain.[15]

Heterosis is thus an important natural phenomenon that can be exploited by properly designed plant breeding programs to increase agricultural yields significantly. The key to the successful commercial production of hybrid seed is sufficient control of the reproduction process. The purity of the inbred strains must be maintained and the parentage of the hybrid seed must be assured. This degree of control of the reproduction processes of hybrid seed is very difficult to achieve on a commercial scale and, until recently, was limited to a few crops that were monoecious, with corn as the leading example. Hybrid corn seed was produced by planting the two parent lines in alternating rows of seed lines and pollinator lines. The tassels were removed from the seed line plants before they were able to shed pollen. This assured that the ears produced by the seed line plants were pollinated only by the pollinator plants, i.e., there was no self-pollination of seed lines. The parentage of the ears produced by the pollinator line plants was also known; they were self-pollinated. For a crop like corn, with distinctly separate male (tassels) and female (silks) flowers, it is possible to detassel (induce male sterility in) the seed lines by hand without too high a cost. However, for crops with small bisexual flowers grouped together, emasculation is difficult, unreliable and expensive. Thus, the greatest strides in creating commercially successful hybrids were made by corn breeders because the plant was so well suited to emasculation through manual operations.

Male sterility, induced manually or otherwise, is fundamental to producing hybrid seed and growing crops with hybrid vigor. While the early hybrid breeding work done with corn was based on detasseling by hand, more reliable approaches to male sterility have been developed in recent years. The most promising of these have been based primarily on sterilizing cytoplasm (extra chromosomal hereditary particles) within the plant; this property causes a malfunction of the male reproductive organs or the spontaneous destruction of the reproductive capability of the pollen. This is a characteristic that, with great patience and effort, can be located in nature or created through crosses between distantly related plants, and can then be bred into varieties to be used as seed lines. The seed lines are thus uniformly and automatically male sterile, without having to resort to expensive and sometimes unreliable manual operations in the field. Cytoplasmic male sterility can make the production of hybrid seed possible for crops where manual sterilization is not feasible on a commercial basis. In China, a source of stable cytoplasmic male sterility seems to have been identified for rice, and work on the development of hybrids is proceeding. Hybrid varieties now dominate not only corn, but also sugar beets and sunflowers in developed countries. Hybrid wheat is being developed, but further work is needed to make it sufficiently superior to the best currently available varieties to cover the extra costs of production. It is generally reckoned that hybrids must be at least 15 percent better than the best open-pollinated varieties to be commercially competitive.[16] Wheat does not yet appear to be at this point, but with another decade or two of research, commercially competitive hybrid wheat production could be a reality.

While hybrid vigor had been observed under experimental conditions by 1910, the commercial application of this phenomenon awaited the development of reliable, cost effective methods of breeding and producing hybrid seeds. By 1920, a technique had been devised that made hybrid corn production commercially feasible. This is the double cross technique that combines four inbred parents. Within 20 years, double cross hybrid seed was being planted on more than half of the corn acreage in the United States, and by 1945 the figure had risen to nearly 100 percent. By 1950, hybrid corn seed was widely used in Europe and quickly came to dominate commercial corn production in France and Italy. This early expansion was based on the use of double crosses, but three-way crosses and, in time, single cross techniques were developed. Tests made as early as 1950 showed single cross hybrids outperforming double crosses and three-way crosses. Results for some of these tests are shown in Table 2-2 where the three types of hybrids are compared with inbreds and open-pollinated varieties.

Table 2-2 illustrates some of the important realities of hybrid seed production. First, the inbreds that are the parents of the hybrid seed have rather low yields, and therefore their production is difficult and costly. Furthermore, the more intensive the inbreeding, the more the vigor of the plant is reduced. Second, the productiveness of the different types of crosses is shown in comparison with contemporary open-pollinated varieties. The average yields of the single crosses are 5 percent more productive than the other hybrids and about 15 percent more productive than the open-pollinated varieties. Tests have established that the margin by which the best single crosses outperform the best double crosses exceeds 15 percent, and selected single crosses are as consistent as the best doubles in terms of yield and other important characteristics. For these reasons, the trend in hybrid corn breeding since 1950 has been to greater use of single cross varieties. By 1980, the share of U.S. corn acreage utilizing single crosses or some modification was approaching 90 percent.[17]

Single crosses are the products of two inbred breeding lines yielding a hybrid seed (A × B). Production of single crosses is relatively expensive because the seed is produced on inbred plants, which as noted above have chronically low yields; therefore, more work and more land

**TABLE 2-2. AVERAGE YIELDS OF HYBRID SEED
IN BUSHELS PER ACRE**

Hybrid Seed Production	Test A	Test B
Inbred lines	40.9	—
Double crosses	98.1	116.5
Three-way crosses	99.7	116.4
Single crosses	104.6	120.6
Open-pollinated	91.3	—

Test A: Jugenheimer 1948–58
Test B: Jugenheimer 1964

Source: Robert W. Jugenheimer, *Corn: Improvement, Seed Production, and Uses* (Melbourne, Florida: Kreiger, 1976), p. 376.

must be utilized to generate the necessary quantity of hybrid seed. When the single cross is used to provide parent seed for breeding double crosses, its low yield level is not as critical because the quantity of seed needed is small compared with the amount required to supply commercial seed markets.

Double crosses incorporate four breeding lines in two single crosses. This produces a hybrid seed (A × B) × (C × D) with more variation than a single or three-way cross. The advantages of double crosses are greater seed yield per female plant and thus cheaper seed. Single cross plants are significantly more vigorous than inbreds and produce about three times as much seed per plant. Moreover, the pollinator plants, being single crosses rather than inbreds, produce more pollen. This makes possible a greater proportion of seed rows in a given field, raising the yield of seed per acre. The vigor of the single cross seed-bearing plants also reduces the impact of adverse environmental conditions. The end product of the double cross has lower yields and more variability in yield and other characteristics than single crosses, but yields are still markedly superior to those of open-pollinated varieties (and if seed cost is an important factor, double crosses can be justified; as a better option than single crosses).

Three-way crosses — (A × B) × (C) — tend to fall between single and double crosses in terms of costs, variability and yield. They are often used where three inbreds exist with good combining characteristics. It should be noted that the inbred (C) is used as the pollen parent so that the vigor and yield characteristics of the single cross are present in the seed producing plants. There are many variations on these three types of varietal crosses, but where hybrid seed is differentiated in trade and production, it is usually categorized as a single, double or three-way cross, as described above.

A summary of the advantages and disadvantages of hybrid seed is in order. The advantages are:

(1) hybrid vigor (heterosis) results in significantly higher yields;

(2) other desirable characteristics can be more easily bred into hybrid seed;

(3) genetic homogeneity assures a predictable growth pattern and a uniform plant;

(4) farmers are assured of the same genetic package and the same performance potential of the hybrid seed from year to year (i.e., hybrids do not "run out," while open-pollinated seed can).

(5) breeders are rewarded in line with the high value they add to crop production over and above the alternative of open-pollinated seed;

(6) farmers do not have to save part of their crop for seed and prepare it for planting the following season.

Three disadvantages are associated with the use of hybrid seed:
(1) hybrid seed is more costly to produce and therefore more
expensive (hybrids, properly chosen, can be superior to open-
pollinated seed even with low inputs, but the extra seed cost
probably will not be repaid);
(2) genetic homogeneity can make the crop more vulnerable
to specific diseases and pests;
(3) farmers are required to buy new seed every growing season
to get the full effect of hybrid vigor in their yields.

The basic reality of hybrid seed production is whether the breeder
can create and produce seed having a sufficiently high yield to offset
the higher costs of hybrid research and production. The alternative
of using open-pollinated seed sets a benchmark for minimum hybrid
superiority, as well as a ceiling on hybrid seed costs, and forces breeders and producers continually to improve their product.

Many hybrid crops are grown today; these include corn, sorghum,
sugar beets, and sunflowers, in addition to several other vegetables,
fruit crops and ornamental plants. The development and improvement
of hybrid seed has enormous potential for raising crop yields over the
next two decades. Research continues to improve corn yields, even
in the United States, and the spread of advanced breeding and production techniques to other areas will result in even more rapid gains
worldwide. A good deal of research has already gone into creating
varieties of rice and wheat with stable male sterility characteristics,
and the next decade could see the introduction of commercially viable
hybrid cotton varieties. Hybrid rice production currently seems more
promising than hybrid wheat, but it is too early to draw a definite conclusion. Hybrid seed production will continue to expand, and hybrid
vigor is one of the most important tools available to plant breeders
in the knowledge era of world agriculture.

The Objectives of Plant Breeding Programs

Before concluding this section on plant breeding, mention should
be made of the diverse objectives of breeding programs whereby breeders attempt to incorporate specific characteristics in plant varieties
to improve their performance. Each crop has its own priority list for
improvements depending on the problems facing the crop and the
needs of its users; however, an overview of breeding objectives will
demonstrate the broad range of challenges facing plant breeders today.
For convenience, these objectives have been grouped into four categories: quantity and quality of product; suitability for diverse growing
conditions; improved physical characteristics; and improvements to
facilitate the breeding and production of seed.

In the first category, quantity and quality of product, the paramount goal is increased yield. This can be achieved in terms of output per acre, per man-hour or per unit of input such as fertilizer. Various approaches are possible, including a greater amount of produce per plant, larger products in terms of size, closer plantings, or even an additional crop per growing season in warmer climates. Dwarf wheat and rice varieties were not desirable in and of themselves, but because they made better use of fertilizer by channeling nutrients to produce more grain. The quality of the product is receiving increasing attention, particularly from a nutritional standpoint. Where possible, breeding programs aim to raise the protein content of crops and to increase the yield of high protein crops, such as beans and peas, to make them more attractive to growers. Another aspect of quality is the appearance and taste of the product. New varieties must be attractive in terms of tastes and preferences to be accepted. With vegetable oils in particular, toxic or unpleasant tasting substances had to be removed through breeding programs before the crop could be widely used. Appearance is particularly important in fruit crops, and breeders must be sensitive to this requirement.

The second category of breeding objectives, suitability for various growing conditions, relates to the general vigor and growth patterns of the plant. The ultimate goal is, of course, high and stable yields for separate varieties adopted to local stress factors. Depending on the conditions in the areas where varieties are to be grown, breeders try to develop different varieties with characteristics suited for growth under specific kinds of stress. Resistance to drought or suitability for given soil conditions are examples of traits that breeders might want to fix in a variety. Plants able to reach maturity in a shorter time or in cooler weather would have an obvious application in northern growing zones. Resistance to pests and diseases is probably the most important breeding objective under the general heading of suitability for local conditions. There are two approaches to this—tolerance and true resistance. Tolerance reduces the probability that the plant will be adversely affected by the disease or pest, while true resistance reduces the effectiveness of the parasite once it comes into contact with the plant. It is not possible to breed a plant with long-term resistance to all pests and diseases; therefore, this battle is ongoing and will continue to be an important challenge facing plant breeders. While these breeding improvements are defensive, i.e., aimed at protecting rather than increasing yield levels, they are nonetheless important. When the disease affects an important staple crop in a sizable area, the need for resistant varieties becomes critical.

The third category, improved physical characteristics of the plant, usually involves secondary goals. Minimum standards must be met to avoid losses in harvesting, transportation, storage, and processing.

Strong stalks are required to prevent wind and rain damage to crops and to facilitate mechanical harvesting. The produce should be relatively uniform in size, shape and composition to reduce wastage in harvesting and processing. Characteristics enabling the crop to be stored and transported without damage are also desirable. Finally, the chemical makeup of the crop may have to be tailored to its end use.[18] Potatoes processed into starch must be different than the various types used for human consumption, while potatoes used for animal feed form a still different variety. In all of these cases, breeders have to create a somewhat different product styled to meet a particular end use.

The fourth category of breeding objectives, characteristics that facilitate the breeding and production of seed, is not immediately obvious to the layman. However, given the difficulty of producing hybrid seed and the power of the hybrid vigor effect, breeders are keenly interested in developing varieties with stable male sterility. This characteristic is invaluable in enabling the efficient production of hybrid seed. Two other important factors in breeding programs are more vigorous inbred parent lines and greater pollen spread from the pollinator lines.

This is by no means an exhaustive listing of desirable characteristics to be incorporated into new varieties, but it does show the complexity of the problems facing breeders. A successful breeding program meets the goal of improvement of a specific characteristic without significant deterioration of any other desirable trait. As an example, a breeding program might aim to better adapt a crop to cool weather without suffering significant loss in yield, disease resistance and so on. Even when improvement of one characteristic is the goal of the breeder, an acceptable level must be maintained for the other desirable traits. Achieving this adds to the time and expense of creating acceptable new varieties; in particular, a large number of experimental varieties must be created, multiplied and tested in the first several years of a breeding program to identify those few new varieties that possess a broad range of desirable characteristics and that have potential for eventual agricultural use.

LEGAL AND ECONOMIC ISSUES

Plant breeding has become a knowledge-intensive industry, heavily dependent upon organized research programs conducted by specialized and experienced breeders. A century ago, plant breeding was done by individual farmers who were improving seed for their own use and benefit. Today, plant breeding has become predominantly a multidisciplinary scientific effort carried out by public agencies, international institutes and private companies that specialize in producing seed. (These agencies, institutes and companies are discussed in more detail in the next chapter.) The international institutes receive the bulk of their funding in the form of grants from private foundations and various public entities. The public agencies are supported predominantly by general tax revenues. The private plant breeders fund their research through the sale of the most successful of the products developed by their research programs. Companies that do not develop improved seed are likely to be overtaken by other companies or the international institutes, lose market share and ultimately be forced out of business: The knowledge embodied in improved seed is the real product of the seed producer, and the force driving private seed producers is competition to produce improved seed, i.e., knowledge. If a company is to cover the costs of production and research and show a return to investors, it must develop new, improved seed that will be purchased by farmers at regular intervals. Failure to create new knowledge in the form of better seed will inevitably lead to declining purchases by farmers. Farmers of course have the choice of buying from other private seed companies, selecting seed from the international institutes and other public suppliers, or saving part of the current year's crop (with open-pollinated varieties).

Three important legal and economic aspects of this process will be explored here: the inherent conflict in pricing and selling intellectual property, i.e., the knowledge embodied in a new seed; the need to ensure competition in supply through a diversity of seed producing companies and organizations; and the time and cost required to develop and bring new varieties into production in today's seed markets. This section will sketch the general principles found to apply to these issues in a large body of economic analysis developed over many years. Chapter 3 will examine the specifics of the seed industry and their relation to the general principles of economic analysis.

Pricing and Selling Knowledge

Improved seed, as with any type of new technological knowledge, differs from ordinary products in one important way: the knowledge embodied in the seed is not depleted through use. Once a plant variety is created, it can be reproduced over and over again without depleting the original store of knowledge in any way. There is a cost involved in actually growing seed, but the cost of producing additional seed is relatively low. Unlike a mineral resource that is depleted or a physical product that is consumed once and for all, the knowledge embodied in the improved seed can be used repeatedly and by many farmers simultaneously without wearing out. This characteristic of knowledge creates a serious obstacle for those who would like to produce knowledge, whether it is in the form of improved seed, original computer software, a new production process, or a mechanical invention. For an investment of money, time and effort in research and development to be worthwhile, an individual or firm must be able to recoup expenses and show a return by selling the results of the research. The problem with knowledge is that once it is produced and becomes available, it can be used by all at little or no cost, and the creator of the knowledge may not get a fair return for his or her contribution. There is an inherent conflict between society's need for knowledge to be available widely and cheaply to achieve its optimal use and the inventor's need to be rewarded for creation of the knowledge. Once the knowledge has been initially developed, the maximum benefit to society results when it is made available at marginal cost, or at what it costs to produce an additional unit. For knowledge, that price is at or near zero. Under such conditions, the tendency is to let others produce and pay for the research and then to co-opt the fruits of the new knowledge by copying and selling it. If such a practice were to become widespread, very few people would produce knowledge.

To ameliorate this problem, patent, copyright, plant variety protection (PVP), and similar laws have been drafted to help writers, inventors, breeders, and other creators of knowledge gain some of the benefits of their "intellectual property." There are, however, two fundamental problems with patent laws in general that have implications for plant breeder protection and other intellectual property laws. The first is that patents are an attempt to reconcile the contradiction noted above, namely, the zero price for knowledge with the need to reward the creators of knowledge for their contribution to society. The second problem is that in practice, individual inventors and firms are increasingly not utilizing their patent "rights" because patents make it too easy for others to replicate the gist of their knowledge (not copy it exactly, thereby violating the patent law) and gain a share of the benefits at very low cost. Once an invention is patented, the information can be purchased at a modest price (relative to the cost of creating the knowledge), modified somewhat and used by others to produce goods and services competing with the original inventor or his or her licensees. In many cases today, individuals and firms feel that the patent system does not provide sufficient protection, and therefore they incorporate their new ideas into their product without putting it on file at the Patent Office for competitors to look over.

In an ideal world, creators of new knowledge would receive a grant commensurate with their contribution as reward for their effort and expense, while the fruits of the knowledge would be distributed freely—or at very low cost. In this way, the maximum worldwide benefit would be derived from all knowledge created. In practice, such a program is not feasible and some sort of second-best strategy must be worked out. The current patent and PVP systems are a reasonable compromise and are the bases on which future improvements would have to be made.

Ensuring Competition in Supply

One line of criticism of the principle of patents and plant variety protection emphasizes "the ability of patent-holding companies to use property rights and the threat of litigation . . . as a means of defining their 'turf' and the opportunity to determine how and who will have access to a patent."[19] Many examples can be cited to demonstrate the point: DuPont built up a strong position in synthetic fibers based on its nylon patents and the General Electric Company (U.S.) used the early patents of Thomas Edison to help support its expansion. In retrospect, one feels that the development of large corporations able to undertake research on a broad front and to achieve economies of scale in the production of goods was inevitable and probably beneficial.

Within the United States, the monopoly-conferring propensities of the patent system have historically been offset by antitrust legislation, and this is an appropriate approach to these types of issues. To use patent laws as a means of reducing monopoly power would be a serious mistake, both because the effectiveness of this approach would be questionable and because it would discourage innovation. If concentration of economic power is the problem, then that should be attacked directly. However, in an era of increasing international competition, very few objective analysts feel that excess concentration is a serious or growing problem. Most countries are finding that to be sufficiently competitive to survive in world markets, firms must be large enough to undertake broad-scale multidisciplinary research programs and to achieve economies of scale in production and marketing. Therefore, the monopoly-conferring potential inherent in the patent system has been considerably diminished in recent decades by the growth of international trade and the resulting increased competition.

Another line of criticism argues that current patent laws afford very little protection because of the relative ease with which the spirit of the patent law can be violated and the benefits of an idea or invention appropriated without violating the letter of the law (or without getting caught at it). The result, as noted above, is that firms and even individuals avoid the patent process, utilize their knowledge in secret and thereby deprive the wider society of the opportunity to work with the new knowledge and to improve upon it. Thus far, this has not been a problem with plant breeders. Where PVP and/or patent laws exist, breeders tend to utilize them and gain legal protection for the knowledge they have created.

Moreover, there is a high degree of willingness to make new varieties available to germplasm banks and to other accredited breeders. A significant diminishment of plant variety protection would have the effect of reducing the exchange of seed knowledge, forcing private breeders to rely more on secrecy for protection and, most important, curtailing the amount of private seed research conducted. To be marketed in most countries, seeds must be registered and certified (described in more detail in later chapters) by undergoing a specified set of test trials to see that the new seed is in fact an improvement on existing varieties. The registration process assures that the important characteristics of new seeds are publicly available and will become generally known. This makes secrecy a less viable option for breeders. Manufacturers do not generally have to register their products and have their performance certified by public authorities; thus, secrecy is an effective way to protect their inventions and innovations.

For plant breeders, however, the combination of no PVP and of registration and certification requirements would be devastating and would dramatically reduce private sector research in only a few years. In the very short term, farmers might be able to buy existing varieties at lower cost, but before very long they would become almost entirely dependent on the research efforts of the public sector and the international institutes for new seed. For many crops, farmers would lose the option of buying seed developed by the private sector, and none of the other sources would be strengthened. The optimum strategy for attaining a secure supply of steadily improving seed is to have diverse sources of germplasm available, to have as many competent research programs as possible, and to have effective competition among several seed producers—private companies, public agencies, foundations, and international institutes—all of which will provide farmers with attractive alternatives for the purchase of their seed. Where there are problems with plant variety protection, some adjustments to the existing system may be in order, but it is not clear whether there is a need for more or less plant variety protection.

The ultimate safeguard for the farmer (for open-pollinated but not hybrid varieties) is the right to save seeds for planting next year's crop. Plant variety protection laws forbid the multiplication and commercial sale of protected seed developed by another breeder. Nevertheless, the laws in most countries specifically permit individual farmers to multiply seed for their own use and, as noted above, this practice is widespread among farmers growing open-pollinated crops. This is one way in which breeders receive less protection than inventors and is certainly an important limitation to a breeder's ability to charge farmers for the full value of improved seed. Farmers growing hybrid crops must buy new seed each year, or run the risk of losing the power of hybrid vigor. Even here, there are limitations to what private companies can charge. In all significant seed markets around the world, a variety of private companies compete to sell seed.

At the international level, intellectual property protection is promoted and coordinated by the World Intellectual Property Organization. This U.N. agency deals with both industrial property—patents, trademarks and industrial designs—and literary and artistic property—books, artistic works, recordings, and films. Within WIPO, an international convention established the Union for the Protection of New Varieties of Plants, commonly referred to as UPOV, in 1961. There are 17 members of UPOV, including Japan, the United States and most Western European countries. Not one of the developing countries is a member, and as of 1983 only Hungary among the centrally planned economies had joined. The commitment undertaken by member nations is to offer protection for plant breeders in conformity with UPOV standards. The substantive aspects of UPOV in general terms provide for:

> (1) national treatment in terms of protection for residents of all other contracting countries;
> (2) right of priority for the breeder who has filed the initial application in another country;
> (3) common substantive rules regarding the form of protection, a minimum number of species to be protected, substance of protection, and length of protection.

The UPOV rules specifically state regarding the substance of protection that "...the breeder's authorization is not required for the use of his (otherwise) protected variety as an initial source of variation for the purpose of creating other varieties...."[20] A breeder's authorization is required for reproduction of the variety for commercial marketing, and this is the extent of the protection offered to plant breeders. UPOV requires that the variety protected must be distinguishable from any other variety, and this can be difficult to do for many crops (cereals and legumes, for example).

Given the natural variability in plants, it is almost impossible to police small-scale infractions of PVP laws in the 17 countries subscribing to UPOV. And in the developing world, where there is no effective plant variety protection, plant breeders can do little to protect against the unauthorized appropriation of new seed for reproduction and sale. The farmers who use the seeds can reproduce them for personal use (or as gifts, for it is commercial marketing that is prohibited) without violating the rights of the breeder; problems of distinguishing one variety from another in a legally satisfactory manner present an enforcement obstacle that does not exist for patents and copyrights; the registration and certification process effectively rules out any chance of relying on secrecy as a means of protecting new varieties; and private plant breeders already face direct competition from institutions that are in large part publicly funded and make their seed available at prices that usually do not cover costs. An unbiased, detailed evaluation of the UPOV system would surely point out areas for improvement. But on the basis of the preceding overview, it is hard to see how a weakening of existing plant breeders' rights would enhance the supply of seed to the world's farmers. The rights currently accorded plant breeders are in fact weak and qualified in nature compared with the rights of inventors and authors.

It is a commonplace of economic analysis that goods, such as knowledge, with significant positive externalities (side effects) are under-produced. Agricultural research is one of the best known of these goods, where the benefits are so diffuse as to be lost or even unknown to the producer. Because the producer cannot capture or even iden-tify the full benefit of the good, it is not produced in sufficiently large quantities, and society as a whole suffers. During the past 25 years, there have been more than 50 studies of the rate of return to agricul-tural research on specific crops in the United States and a dozen other countries. The overwhelming evidence is that the real rate of return (after inflation) is in excess of 15 percent in the United States for every crop studied. Real return to agricultural research in the other (mainly developing) countries varied widely, but, on average, these returns were well in excess of 15 percent.[21] There are very few sectors in which the rates of return to investment have been this high for more than a quarter of a century, yet real research expenditures by the U.S. federal government have stagnated since 1965. Public support for agricultural research by state and other nonfederal agencies has done better, roughly doubling between 1960 and 1980, and is approaching the fed-eral government in terms of overall level of effort.[22] On balance, how-ever, public agricultural research in the United States has lagged behind recognized needs despite evidence of very high rates of return. The reasons for this, as suggested above, lie in the difficulty of identifying and capturing the benefits for the agencies funding the research. The effects of research inevitably spill over to provide benefits for indi-viduals and organizations that do not fund the research.

· During the 1970s, over one-third of the U.S. wheat acreage planted in varieties created by public agencies used products developed in another state.[23] This implies that much of the effect of a given state's research program is lost to competing farmers from other states who do not contribute taxes to support the program. In addition, smaller states seem to have agricultural research programs that are more than proportionately smaller than those of the larger states. A smaller state will capture less of the benefits it generates and, conversely, finds it easier to be a free rider benefiting from the larger research programs of nearby larger states.

A second spillover effect takes place internationally when a share of the benefits of agricultural research accrues to countries that do not contribute to the tax base supporting the system. Two among thou-sands of such examples are soybeans in Brazil and corn in Italy. A further consideration is the rapid transfer of the benefits of agricultural research to consumers in the form of lower prices for agricultural products. Where markets are characterized by slow growth in demand — as food markets in the developed countries are — most of the benefits from productivity gains result in larger output at lower prices. Again, the benefits of agricultural research tend to escape from the producers and the producing regions that support and fund the research programs. A large share of these benefits goes to consumers who are unaware of the connection between agricultural research and cheap, plentiful food. The result of these spillovers is the chronic tendency to underfund and underproduce agricultural research, despite all the evidence of the high rates of return to this type of investment. Given this well documented, chronic tendency, it would be unwise, to say the least, to force cutbacks in private research by undermining plant breeders' rights.

The Time and Cost of Plant Breeding

This chapter will conclude with a brief description of the time, expense and risk required to bring a new plant variety through development, certification, production, and ultimately to acceptance by farmers. It is generally agreed that the breeding time required to produce a new variety is 10 to 15 years, depending upon the crop and the difficulty of the objectives of the breeding program. Under special circumstances, this can be accelerated by doubling up the growing season, that is, by running tests or multiplying seed in the tropics, or the Southern Hemisphere, during the northern winter.

The first step of any program is to recognize what needs to be done and to establish the objectives of the breeding program. Breeding objectives fall into five broad categories — yield, quality, resistance, adaptation to local conditions, and cost of production. Ratios vary, but an efficient breeder should aim to start with an average of between 3,000 and 5,000 viable crosses for each successful variety. Skill and luck are of course factors, and beginning with a larger number of crosses is no guarantee of success. Progeny selection, testing and evaluation proceeds for several years until a few individual plants with promise are selected. Once a variety has shown its strengths in terms of the objectives of the breeding program, which takes 8 to 12 years, the breeder will move to evaluate the market potential, apply for plant variety protection and begin multiplying seed from individual plants. In countries requiring certification trials, that process will begin, with the goal of introducing an attractive new variety to the market as soon as possible.

The approximate time frame for each of these stages for a hybrid and open-pollinated variety is presented in Table 2-3. Many of the stages overlap and can be carried out simultaneously. Others, however,

**TABLE 2-3. STAGES AND TIME REQUIRED IN
PLANT BREEDING
(Time in Years*)**

Hybrid	Stage	Open-Pollinated
0	Recognition	0
4-5	Parent line preparation	0-2
5-6	Initial crosses	0-3
6-10	Progeny selection	3-11
7-12	Crop evaluation	5-12
7-15	Testing the variety	6-15
8-12	Determination of a new variety	7-13
8-13	Market evaluation	8-14
8-14	Application for plant variety protection	9-14
9-14	Multiplication from individual plants or ears	9-15
9-14	Certification	9-15
10-15	Market introduction	10-17
12-18	Market acceptance	12-19
13-19	Market growth	13-20
20-25	Obsolescence	20-25

*The time frames shown are indicative and can be accelerated
by doubling growing seasons and other methods that entail
additional expense.

Source: Interviews with plant breeders.

are sequential and limit the degree to which any breeding program
can be shortened. It is worth noting that breeding programs for hybrids
tend to require more time working on the inbred parent lines, whereas
breeding programs for open-pollinated varieties tend to take more time
with progeny selection. This difference is as expected, given the importance of getting the inbred parent lines properly established in hybrid breeding.

The figures in Table 2-3 indicate the range of years in which the
particular activity would be carried out, depending on a variety of factors influencing both the overall program and the specific stage. The
two most important factors are probably the urgency of the need for
the new variety and the breeding skill and management capabilities
of the key people conducting the program. The timetables noted in
Table 2-3 do not represent the time required to create new varieties,
such as the high yielding dwarf wheat and rice lines introduced in the

1960s. These programs take even more time—perhaps an additional 5 to 10 years—in terms of preparing the parent lines, selecting the progeny and evaluating the research.

Another perspective on the time factor involved in bringing new varieties forward is given by Table 2-4, which notes the average number of years required to breed several fruit, vegetable and cereal crops. There is considerable variation with cauliflower and squash at the high end and corn and safflower at the low end. The time spent preparing the parent lines for hybrids is presumably not included and that would change the figures somewhat. A note of caution regarding the data for individual crops in Table 2-4 is in order, as many of the individual samples are too small to draw definitive conclusions. Note also that this table takes the time frame out only to the application for PVP, so three to six years still remain before most of the varieties sampled would be in the marketplace.

In terms of the financial expense of a breeding program, published estimates indicate that the research for a new variety costs in the range of $1.5 million to $3.0 million.[24] The cost of a modest breeding program, including a chief breeder, a staff of three or four, an experimental station, equipment, and land, is about $250,000 annually in the United States. Costs overseas will vary depending on the current exchange rate; but if the $250,000 figure is multiplied by the 6 to 10 years it takes to conduct the research for a new variety, the result is $1.5 million to $2.5 million. In addition are the costs of producing and marketing the seed and administering the whole operation. Larger breeding programs are probably able to achieve some economies of scale by working on several varieties simultaneously, but administration costs might also be higher. Therefore, an estimate of $2.0 million to $2.5 million for the breeding research for a new variety in the United States is probably accurate.

As breeding becomes a more complex and multidisciplinary process, costs can be expected to rise. Accelerating the time frame by doubling up on the growing season requires breeding stations in two distant locations and implies the use of jet cargo planes to move seed and plant matter back and forth. This raises costs and complicates the management of the breeding program, but ought to increase the rate of output of new varieties within a few years. The introduction of computer monitoring of many plant crosses for a multitude of characteristics should allow breeding teams to generate more marketable varieties, but also implies greater equipment costs and continuing need for botanically literate computer programmers. There are also long-term research programs. Attempts to develop hybrid varieties of wheat, rice and other potential hybrid crops via cytoplasmic male sterility have absorbed research funds for 25 years and are not yet profitable. Finally, the potential of biotechnology introduces another level of com-

TABLE 2-4. RESEARCH TIME REQUIRED FOR DEVELOPING NEW VARIETIES

	Years			Number Varieties*	
Crop	Cross to Date of Determination	Date of Determination to Application	Total	Cross to Date of Determination	Date of Determination to Application
Barley	7.0	3.4	10.4	9	10
Beans	8.0	3.3	11.3	16	32
Cauliflower	11.0	7.5	18.5	1	4
Corn	5.5	2.0	7.5	4	6
Cotton	8.0	4.2	12.2	27	57
Lettuce	7.0	2.6	9.6	22	34
Oats	8.8	2.1	10.9	13	16
Onion	9.0	2.9	11.9	11	14
Peas	7.0	4.0	11.0	24	34
Rice	6.0	2.8	8.8	5	12
Safflower	6.0	1.7	7.7	4	5
Soybeans	6.2	3.0	9.2	64	75
Squash	11.0	3.7	14.7	1	3
Tobacco	8.5	2.6	11.1	11	14
Tomato	8.3	1.4	9.7	3	9
Watermelon	8.5	5.0	13.5	2	10
Wheat	8.0	2.8	10.8	36	56
Average or total	7.9	3.2	11.1	253	391

*Applicants are required to list the date of variety determination and date of application when submitting protection applications. They are not required to list the date the cross was made. It is for this reason there are fewer varieties listed in the "Cross to Date of Determination" column.

Source: Reproduced by permission of Asgrow Seed Company, Kalamazoo, Mich. 49001, published by Asgrow Seed Company, June 27, 1985, in "A Chronicle of Plant Variety Protection 1983 Update," p. 25.

plication that plant breeders will almost certainly have to master to remain competitive. All these changes improve the potential of plant breeding and during the coming 10 or 15 years should generate some notable successes in terms of new varieties; in the short term, however, they are raising the costs and the risks of successful plant breeding tremendously.

Risks are an inherent part of the economics of plant breeding. Five types deserve mention—technical, registration, marketing, environmental, and political/legal risks. There is always a risk in plant breeding that none of the varieties developed will meet the objectives. As noted above, a large number of crosses and the maintenance of a large number of lines through the early selections make it more likely that a few acceptable varieties can be found, but there is no real guarantee. It is the ability to solve the technical problems that defines a good breeder, and this type of risk is well understood and accepted by all who breed plants for a living. After the technical problems have been overcome, the plant breeder faces the risk of not having the new variety approved in official trials and registered as an acceptable or recommended new variety. In many countries, including Canada and the majority of Western European nations, the state prohibits sale of varieties that are not registered and on the official list. In these countries, less than 10 percent of the new varieties tested are eventually listed.[25] This, then, is a significant risk for breeders—even after the technical problems are solved, 9 out of 10 new varieties are rejected in official trials in many major markets.

There is still a risk after a variety is able to be marketed that it will not be commercially successful. Success in official trials is no guarantee of acceptance by farmers, and it is a minority of about 25 percent of the varieties reaching the market that achieve market acceptance. In countries without official trials, a larger proportion of products reach the market, but the failure rate is correspondingly higher. At this stage, the failure of a variety generally occurs because the market has changed, better improved varieties have been introduced, or the original objectives of the breeding program set 10 to 15 years ago were not good enough. This is the competitive risk facing seed producers whereby farmers impose their preferences on the seed supply industry and reward the most successful plant breeders.

The fourth type of risk involves the vulnerability of a new seed to climate, disease, pest, or other environmental conditions encountered after the new variety becomes widely used. It is not always possible to anticipate the stresses a variety will face once it is grown as a crop, and new strains of disease or insects can develop. Thus, even after a variety has been accepted by the market, it may demonstrate a vulnerability to a new or unanticipated environmental factor that dramatically reduces its usefulness to farmers. Although risk of this type of failure is very low, it can abruptly end the commercial life of an otherwise successful plant variety.

Finally, there is the political/legal risk that a sizable part of the expected gain from a new variety will be lost because of inadequate plant variety protection (this is particularly true for the self-pollinated crops). Where PVP laws are lacking, a successful breeder may have little or no recourse if that seed is multiplied and sold by others. And even in countries with protection, it is not possible or practical to police minor violations. Under such conditions, it is unlikely that a private seed company will direct a sizable research effort to meet the seed requirements of countries that do not have a history of protecting plant breeders. In those countries, the political/legal risk is too great; a company can still fail even when all the other risks have been overcome and the full investment in time and money has been made.

The risks described above are the most important aspects of the cost of developing new plant varieties. The time factor is known and is reasonably predictable, as is the average expense. However, as the breeding process is becoming more complicated, it also is becoming more expensive. On average, time and expense are manageable problems because they can be covered by the generally high average margins generated by sales of the really successful varieties. The problem arises when a company faces a big risk and loses. A large company can spread risk around many products and expect to survive, but a small or medium-sized seed company is likely to be either thriving or struggling. With even a few very successful varieties, a company can do extremely well; but with rising research costs and high risks in trying to replace current successes with future successes, the fortunes of a company can turn down very quickly. The next chapter looks at the world seed industry and the dramatic restructuring that has taken place during the past decade. Large companies, often chemical and pharmaceutical corporations, have absorbed many of the independent seed companies in Europe and the United States. This trend has been due to the attraction of the high average profits of the seed trade and the enormous promise of biotechnology for future seed production, but also to the increasing risks facing seed producers. The large companies are better situated to accept such risks and to fund the rising cost of research, but only time and their performance will tell if they can achieve their potential and succeed as plant breeders and seed producers.

FOOTNOTES, CHAPTER 2

1. On this process, see Jose T. Esquinas-Alcazar, *Los Recursos Fito Geneticos, Una Inversion Segura Para El Futuro*, Ministerio Agricultura, Madrid, Spain, 1982, pp. 9 and 10.
2. J. Sneep and A.J.T. Hendricksen; *Plant Breeding Perspectives* (Wageningen, Netherlands: Center for Agricultural Publishing and Documentation, 1979), p. 84.
3. Esquinas-Alcazar, pp. 12–15.
4. Andrew M. Greeley, *That Most Distressful Nation* (Chicago: Quadrangle Books, 1972), pp. 34–35.
5. Sneep, p. 92.
6. Ibid.
7. Ibid., p. 93.
8. Ibid., p. 94.
9. See *Seeds for a Hungry World: The Roles and Rights of Modern Plant Breeders*, by the Canadian Seed Trade Association (O'Haura, Canada).
10. J.H. Longquist, "A Modification of Ear-to-Row Procedures for the Improvement of Maize Populations," *Crop Science* (1964), pp. 227–228.
11. Sneep, p. 191.
12. Ibid., p. 106.
13. Robert W. Jugenheimer, *Corn: Improvement, Seed Production, and Uses* (Melbourne, Florida: Kreiger, 1976), p. 75.
14. William Brown as quoted in Sneep, p. 191.
15. Jugenheimer, p. 88 for corn, and Sneep, p. 173 for rice.
16. Sneep, p. 172.
17. Jugenheimer, p. 382.
18. Sneep, p. 191.
19. Pat Ray Mooney, "The Law of the Seed," in *Development Dialogue*, Vol. 1–2 (Uppsala, Sweden: Dag Hammarskjold Foundation, 1983), p. 137.
20. *WIPO, General Information*, WIPO Publication No. 400 (E), p. 36, March 1983.
21. Vernon Ruttan and W. Burt Sundquist, "Agricultural Research as an Investment: Past Experience and Future Opportunities," Report of the 1982 Plant Breeding Research Forum (Pioneer Hi-Bred, 1983), p. 60.
22. Ibid., pp. 90–92.
23. Ibid., p. 72.
24. *Feeding the 5000 Million* (Amsterdam: ASSINSEL, 1981), p. 10.
25. Ibid.

The Seed Industry

3

The seed industry is part of the genetic supply industry that historically has included seed and livestock breeding for agriculture. In the future, the genetic supply industry will be much broader, involving the production and supply of a whole spectrum of new, genetically based products to industry as well as agriculture. The seed industry is in the midst of profound changes, as this chapter will make evident, and it is not possible to comprehend what is under way unless the seed industry is viewed as an entre to one of the most dynamic growth sectors expected in the next quarter century—the genetic supply industry.

In the 1970s, it became clear that basic research breakthroughs had the potential to change radically and accelerate the process of improving plant varieties. The potential is now also present to generate profoundly modified organisms with the capacity to alter and/or accelerate chemical, biological and other processes in agriculture and industry. It was found that the genes, which maintain and transmit the inherited characteristics of a species, are composed of combinations of nucleotides, the basis of deoxyribonucleic acid which is located in the chromosomes of the cell nuclei. Basic research into the structure of DNA and its roles in heredity created the potential for directly altering the genetic makeup of a species, rather than using the time-consuming procedure of selective breeding. This potential for producing new types of organisms through bioengineering has attracted many new firms to the genetic supply industry and has contributed to a dramatic restructuring of the seed industry.

AN OVERVIEW OF THE INDUSTRY'S RECENT EVOLUTION

The seed industry includes institutions, organizations and business firms that perform four different types of activities: basic research, applied research, seed production, and marketing and distribution. The line differentiating basic research and applied research is hard to define, but worth drawing. Treating basic research as a separate activity focuses on the contribution of the research programs in plant breeding conducted by state, national and international institutes. Long-term basic research has traditionally been performed by public sector entities, such as the Department of Agriculture and state university agricultural research programs in the United States, the National Institute for Research in Agronomy (INRA) in France, and some of the International Agricultural Research Centers (IARC) in the developing world. A few of the products of these research programs are directly marketable, but in the advanced countries, varieties developed by the public sector are used primarily by private firms in applied research programs. Thus, while basic and applied research overlap somewhat, a complementarity has developed over time whereby publicly supported institutes focus on basic research and private firms on applied research, production and marketing, and distribution. In developing countries, private firms are not as strong because of various constraints, and public agencies are commonly entrusted with applied research, production and marketing.

In the past, there was more differentiation of firms by activity. Many small seed companies did not have research programs and utilized the research efforts of the public sector in producing and distributing seed for a small range of crops over a specific geographic area. In the United States, foundation (basic) seed companies were the chief agents in making the initial increase of publicly developed varieties. Seed firms have traditionally been small, independent and often family-owned. The largest and most successful have had integrated activities, including applied research, production and marketing, but have been low on any listing of large U.S. or international business corporations such as the Fortune 500. During the 1970s, this structure was altered fundamentally by a substantial number of mergers and takeovers. With only a few exceptions, the most successful seed companies in Europe and North America were taken over by much larger firms, with their base of business activity in the pharmaceutical, chemical, petrochemical, or food processing sectors. In North America, at least 100 firms have been merged or acquired since the early 1970s. The objective of these takeovers seems to have been the establishment of integrated genetic supply firms that are positioned to participate in the industry's projected growth. This seems to be confirmed by the expansion in research that typically followed the acquisitions. In general, the firms acquired conducted little research, focusing instead on production and marketing and distribution; after the takeover, the parent firm then began to build up research, emphasizing biogenetic work more than traditional breeding programs. Perhaps the leading example of this is Agrigenetics, which combined 12 small seed firms in several crops, raised $55 million for research, then contracted with several universities and other organizations to undertake biogenetic research.[1]

At the same time that fundamental structural changes are taking place among the firms in the seed industry, important changes are under way among the national and international plant research organizations. There is great concern that funding for public sector plant research, particularly in the United States, is not keeping pace with the need for basic research or for the maintenance of a diverse germplasm resource base for both public and private breeders. The expansion of private sector research programs coupled with a stagnation in the level of real (after inflation) federal funding for agricultural research may be pulling scientists away from publicly funded basic research programs. At the very least, this sector is not growing, and at a time when doubts exist about the adequacy of the current level of effort. It is unlikely that public sector institutions will be able to participate appropriately in basic biogenetic research in the coming "molecular revolution"; public agencies may not be able to continue their traditional roles in plant breeding and germplasm maintenance without increased funding.

The international plant research centers also seem to be facing a difficult period. There is dissatisfaction among developing countries with the operation of the international agricultural research programs. The problem currently centers on the International Board for Plant Genetic Resources (IBPGR), but the same complaint may be applicable to other parts of the overall program, including the International Agricultural Research Centers. Briefly, there seems to be a desire for increased control of plant varieties and genetic research by Third World countries; they believe the consortium of developed country governments, foundations and international and private organizations that fund and dominate the international agricultural research programs does not adequately represent Third World interests. In particular, some criticize the current institutions for allowing excessive benefits of genetic material from the zones of origin (largely in the developing world) to accrue to multinational seed companies. The developed countries, on the other hand, believe the existing system is reasonably effective and are reluctant to make major changes. This fundamental disagreement is reflected by events that have occurred at recent general meetings of the U.N. Food and Agriculture Organization (FAO).

All three components of the seed industry—private companies, public organizations and international institutes—are in a period of greater transition and turmoil than they faced in the 1960s and 1970s. The opportunities and challenges are great, but the problems and dangers are very real and very apparent. This chapter will cover in detail the background, current trends and prospects for each component of the seed industry and will conclude with a discussion of plant breeders' rights. The strategic role of the seed industry in maintaining a favorable world food balance makes it imperative that current problems be resolved and the productivity of the seed industry be protected and enhanced. To achieve less is to risk a faltering in international food production and the enormous consequences that would ensue.

THE INTERNATIONAL AGRICULTURAL RESEARCH SYSTEM

The origin of the current network of International Agricultural Research Centers was a jointly sponsored program set up by the Mexican Department of Agriculture and the Rockefeller Foundation in 1943 to work on wheat and corn. Programs continued on an ad hoc basis until 1960 when the Rockefeller and Ford Foundations established the International Rice Research Institute (IRRI) in the Philippines. This endeavor was the source of the basic research on the high yielding rice varieties. In 1966, the Mexican programs were formally established as the International Center for the Improvement of Corn and Wheat (CIMMYT), and this center was responsible for the basic research that developed the high yielding wheat varieties. In 1971, chiefly at the instigation of the World Bank, the Consultative Group on International Agricultural Research was formed to address agricultural problems and raise food production in the Third World. CGIAR is an association of governments, foundations and international and private organizations and includes the World Bank, the FAO, the Rockefeller and Ford Foundations, and other private foundations such as the Kellogg Foundation. It is primarily intended as a conduit for funding and a coordinating body, with the chairman and staff provided by the World Bank. The scope of CGIAR's activity takes in food crops, livestock programs, plant genetic resources, and food policy in general. More than a dozen international agricultural centers and programs, including IRRI, CIMMYT and IBPGR, are dependent on CGIAR for the bulk of their funding (see Table 3-1 for a listing of the research centers). The system also attempts to help developing countries improve their research techniques and make their applied programs more effective. During the 1970s, CGIAR funding grew dramatically from about $21 million in 1970 to $136 million in 1980. Funding has leveled off somewhat in recent years, as inflation has pushed up costs faster than the growth of funding, forcing curtailment in various activities. Nevertheless, total funding reached $174 million in 1985.[2]

CGIAR then, is the financial and oversight centerpiece of the existing system of international agricultural research. In the view of the donor countries, this system has worked well and has justified the resources funneled into it over its history. By comparison, the previous system with the FAO at the center was less than satisfactory. Although not officially acknowledged, many felt that FAO's efforts were sometimes marked by the same dominance of politics, cronyism, bureaucratic ineptitude, high overhead costs, and lack of results that seem to have characterized certain other U.N. agencies. Currently, however, the majority of the donor countries and agencies believe that CGIAR is succeeding in terms of its goals and responsibilities and thus do not support significant changes.

TABLE 3-1. THE INTERNATIONAL AGRICULTURAL RESEARCH INSTITUTE NETWORK

Center	Program	Year Founded	Headquarters	Agroclimatic Area Served
International Rice Research Institute	Rice, multiple cropping	1960	Philippines	Rainfed and irrigated areas —subtropical/tropical
International Center for the Improvement of Corn and Wheat	Wheat, corn, barley, triticale	1966	Mexico	Rainfed and irrigated areas —temperate/tropical
International Institute of Tropical Agriculture	Corn, rice, cowpeas, soya beans, lima beans, cassava, yams, sweet potatoes, and farming systems	1968	Nigeria	Rainfed and irrigated areas—lowland tropics
International Center of Tropical Agriculture	Beans, cassava, beef and forages, corn, rice, and swine	1969	Colombia	Rainfed and irrigated tropics—1000 meters to sea level
International Potato Center	Potatoes	1972	Peru	Rainfed and irrigated areas—temperate-to-tropical
International Crops Research Institute for the Semi-Arid Tropics	Sorghum, millet, groundnuts, chick-peas, pigeon peas	1972	India	Semiarid tropics
International Laboratory for Research on Animal Diseases	Blood diseases of cattle	1974	Kenya	Mainly semiarid tropics
International Livestock Center for Africa	Cattle production	1974	Ethiopia	Humid to dry tropics
International Center for Agricultural Research in Dry Areas	Wheat, barley, lentils, broad beans, oil-seeds, cotton, and sheep farming	1976	Lebanon, Syria Iran	Mediterranean

Source: Reprinted by permission of *To Feed This World: The Challenge and Strategy*, by S. Wortman and R.W. Cummings, Jr. (Baltimore: Johns Hopkins University Press, 1978).

On the other hand, critics of the CGIAR, some from the developing countries, argue that the subtropical and tropical regions, as the origin of most of the genetic material supporting world agriculture today, have not benefited appropriately. They contend that much of this genetic material was removed without the countries' permission and some even taken in violation of existing laws. Plant genetic resources from the Third World have sometimes been used to develop commercial seed lines that are then sold back to developing countries.[3] To put this argument into perspective, a brief look at the geographic movement of plant varieties is in order. Since antiquity, the migration of seeds and plants has been the means by which plant varieties have spread to new regions. Seeds usually spread by natural means—birds, streams and the wind—or migrating peoples carried seeds with them. In some cases, new species of plants became the staple foodstuff in their new habitat; potatoes and tomatoes, for example, were brought to Europe from the New World. The acquisition of potentially valuable plants from varied and distant locations is as old as organized agriculture. In ancient times, the Egyptians made an effort to acquire new species of flora whenever possible. Food crops, particularly grains where the edible portion was the seed itself, spread widely and naturally to suitable neighboring regions; other species, including spices, beverage bases and medicinal plants, did not migrate naturally, but were very valuable commodities in commerce. Attempts were made to embargo the movement of these plants' seed; but the small size of seeds and the ingeniousness of smugglers always frustrated would-be monopolists, and plants slowly and steadily became more widely available.

Some argue that an element of exploitation existed in the movement of plant species from their place of origin to other suitable locations for propagation. Crops most frequently mentioned are coffee, tea, cocoa, rubber, bananas, sugar cane, spices, and cotton.[4] With the exception of cotton, none of these currently are important crops in Europe or North America. They are examples of transfers between various regions in the developing world: coffee moved from Ethiopia to Brazil; rubber from Brazil to Malaysia; sugar cane from Southeast Asia to Brazil and the Caribbean. None of the major staple crops of the developed countries are included except for potatoes and soybeans, because they had migrated from their zones of origin at a very early stage in the development of civilization and farming, i.e., before the arrival of the Europeans during the Age of Discovery. The earliest settlers in all regions of North America found corn being cultivated by the native Indians. Varieties of corn probably have been spreading northward since the last ice age, and it is certain varieties from this naturally selected pool of germplasm that dominate breeding programs in the United States today. This is also true for most other plant varieties used as commercial seed in the developed countries.[5] Wheat and barley, for example, moved into Europe from the Near East at the dawn of civilization. "The facts are that there are very few varieties of plants in commercial use in developed countries whose genetic background includes any significant amount of recently added exotic germplasm."[6] Certainly these varieties originated many millenniums ago in regions that are now geographically within the developing world; however, the emigration of suitable varieties was well advanced before humans developed a conscious memory. To imply that some sort of exploitation occurred is to misunderstand the genetic basis of staple crops in the developed world. Soybeans went from China to America and corn from America to China during the Age of Discovery; sorghum went from Africa to China much earlier. The web of exchange is very tangled and goes back without break to the beginning of farming. The noteworthy migrations of germplasm in recent years, in fact, have been to and within the developing countries; examples are North American soybean varieties introduced in Brazil and Mexican wheat strains used for breeding in India and Pakistan.

It is tragic that this misperception has been created because the losers will be the developing countries, and there is evidence that a good deal of damage has already been done. According to one report, "Many in the Third World are outraged by the argument that they freely supplied Western scientists with most of the seed stock that was used to create improvements."[7] In reaction, seed wars are starting with countries possessing the richest endowment of certain plant varieties refusing to let seed leave the country. As a result, Central America suffers from coffee tree blight because Ethiopia will not allow its coffee seed to leave the country. Similarly, India is embargoing its black pepper seed; Southeast Asian countries are withholding certain types of fruit; Ecuador is holding back cocoa; and Taiwan is restricting sugar cane. In each case, the crop is not grown in northern agricultural regions and the damage is being done to Third World agriculture. In time, the seed will get out because embargoes inevitably fail; meanwhile, farmers in the developing countries are the big losers.

National and international seed banks and research agencies supply free samples from their collection and distribute tens of thousands of varieties at little or no cost every year from these organizations. However, many proprietary lines, especially the inbred parent lines used to produce hybrid seed, are closely held by their breeders and generally unavailable for use by others. The sale of hybrids represents a release of germplasm, and these commercially released varieties are available for interseeding and the development of new inbreds. There is also a problem in exchanging seed with Eastern bloc countries. Requests are sometimes ignored, or they are returned with improper labeling or inadequate descriptions of the sample sent back. A great deal of secrecy surrounds the status of Russian seed conservation programs. Because of the early work of Vavilov in the 1920s, the Soviet Union began with the best collection of seed in the world. Whether these were properly attended, preserved and grown out through the years — including the difficult years of World War II — is not known. The problems of getting prompt, accurate responses to seed requests and the general secrecy surrounding Russian seed collections suggest a significant part of Vavilov's original collection has been lost.

Americans are much more open about their successes and problems. On the plus side, in 1980 the U.S. government system of gene banks based in Beltsville, Maryland, outside Washington D.C., distributed over 250,000 free samples of seed in response to requests from 50 countries out of their collection of over 225,000 accessions.[8] On the negative side, there is a chronic inadequacy of funds for the U.S. gene conservation programs, including the long-term storage center, the National Seeds Storage Laboratory, in Fort Collins, Colorado. More funds are to be allocated to these programs, but this still falls short of what is required to bring the U.S. conservation system up to desirable levels. The problems facing the U.S. program are typical of difficulties facing similar programs in most other countries.

Seed conservation is simply not a high priority item when it comes to allocating funds from public budgets. As noted in Chapter 2, there is a desperate need to increase efforts to conserve diverse sources of germplasm, yet the existing funding is not adequate to maintain properly the acquisitions currently in hand. Increased funding for genetic conservation should have top priority within the international seed community. A well organized, coordinated effort should be made to get appropriation increases, and any proposals for change that do not quickly lead to an increase in the resources available should be deferred until the top priority problem — insufficient funding — has been dealt with satisfactorily. With the potential for an outbreak of seed wars among the developing countries, the existing gene banks become even more important as sources of diverse germplasm. Any action to weaken this system would do the initial and strongest damage to the developing countries which are most dependent on public gene banks. The effects will be felt more immediately in terms of lower incomes for Third World farmers, loss of foreign exchange and, in the case of staple crops, a poorer diet for the average person. Irresponsible policies based on faulty analyses will ultimately lead to a deterioration in the food balance and increased hunger in the developing world.

The focus of this debate has been the International Board for Plant Genetic Resources, formed in 1974 by the CGIAR, the FAO and the U.N. Environmental Program (UNEP). The IBPGR was established to accelerate the conservation and preservation of genetic resources in the face of heavy pressure from the spread of the genetically narrow high yielding varieties. As described in the previous chapter, there was growing concern that natural varieties were being lost forever. The existing organization (FAO's Crop Ecology Unit) seemed unable to expand its activities quickly and efficiently, and a new initiative, specifically the IBPGR, was needed. Funding began at $1 million in 1974 and grew to $4 million by 1983, but IBPGR has remained in an anomalous position. It is housed with FAO in Rome, yet the CGIAR provides the bulk of its funds and controls overall policy. Under the current set of procedures, the IBPGR works with the International Agricultural Research Centers and national governments to collect, classify, preserve, and ultimately distribute plant germplasm. While most of the natural plant germplasm is in the zones of origin in Third World countries, many of the gene banks and most of the genetic research programs are in the developed countries. (It should be noted that most of CGIAR's gene banks are located in developing countries.) Hence, to some of its critics the current system appears to be a device to channel germplasm out of the developing countries and their control and into the hands of commercial breeders in Europe and North America. This dissatisfaction was exacerbated in the 1970s by the enactment of plant variety protection laws that strengthened plant breeders' rights in the developed countries. The developing countries felt that seed companies thus had access to newly discovered native varieties for breeding purposes at low cost, while selling significant volumes of the products of their breeding programs at high cost—and had increasing legal protection against those who would reproduce the seed for sale. The views of the developing countries surfaced officially in 1981 at the general meeting of the FAO and culminated at the November 1983 session. At this meeting, a resolution was passed to formalize a new system (to be discussed below) giving the developing countries greater control over the IBPGR and the world's plant germplasm preservation system. In the voting, 144 countries recorded for changing the system, while 8, including the United States and several European countries, opposed.

By now the matter has become so politicized that sorting out the issues and conflicts is almost impossible. On the one hand, some believe that there is merit in the view of the developing world, and perhaps there should be payment of royalty fees or some compensation for the contribution of Third World genetic resources to commercially valuable varieties. But it would be a real challenge to establish a basis for quantifying this fee. On the other hand, the position of the Third World ignores the substantial benefit that developing countries have received from international agricultural research. At little or no cost to them, food production in the Third World has increased dramatically, as noted in Chapter 1. Even netting out the cost of imported fertilizer and equipment, the benefits of the existing system far outweigh the costs that developing countries have paid. A second point is that sales of seed by private companies in developing countries are neither particularly large nor profitable, thus, it is hard to understand the basis for the accusation that seed is being sold in developing countries at unacceptably high prices. In most Third World countries, the government, through its agricultural agencies or enterprises, is the leading seed supplier in the market. Various nonprofit groups, including the international agricultural centers, are usually in the background supporting the government. Further, new seed must be certified by government agencies and testing in field trials before being allowed on the market. Farmers in developing countries tend to be extremely resistant to change and reluctant to adapt new techniques and approaches. In such a situation, the private firm has to have a certifiably superior product to have a chance of breaking into the market, let alone be able to charge a higher price.

Developing countries further contend that, even if their germplasm is not reconstituted and re-exported to them with high markups, it is used to great profit by farmers in the developed countries. This is true and may justify a type of royalty arrangement, but the argument overlooks the fact that excess supplies of agricultural production return to developing countries as food imports (which are often subsidized). Thus, the food importing developing countries benefit considerably from the low world food prices resulting from the productivity of farmers in North America, Australia and Europe. The benefits of germplasm coming from the zones of origin are widely dispersed and do not reside exclusively with private seed companies or with farmers in the developed countries.

From the viewpoint of economic efficiency, there is only weak support at best for the position that higher prices in the form of royalty payments should be paid for access to the germplasm in the zones of origin. The relevant line of argument is the same as that for patent protection, except that the costs of creating the newly discovered variety are zero. The costs of breeding a new variety are significant, but, by definition, the natural varieties already exist, and no incentive is needed to encourage people to produce them. However, funds are needed to cover the costs of searching out native varieties, preserving them and making them widely available at low cost. Just as with knowledge and the patent argument, once a new native variety is found, the benefits are maximized by the widest possible distribution and use. And because there is no cost of producing the variety, unlike the product of a breeding program, there is no efficiency argument for charging those who benefit from the newly discovered variety and rewarding those who happen to live in the region where the native variety was discovered. It is not an issue of rewarding "private interests," as some would characterize it, but of rewarding productive effort. The developing country argument can be turned around: "we have genetic material and you have knowledge and technical skills; let's combine them and fully distribute the results." The problem here is that the developing countries need to contribute little, if any, productive effort to the partnership. They do not have to work to acquire the genetic diversity they possess, whereas breeding programs require years of effort and large commitments of resources. The level of payments justified on the basis of efficiency is that which would adequately fund the preservation activities of the IBPGR to meet the costs of searching out new native varieties and bringing their germplasm into use in breeding programs.

While the above argument is correct from an efficiency point of view, and has very likely been part of the intellectual justification for the existing system, it may not suffice in the future. Some argue that there may be justification for royalty payments; certainly a compromise needs to be worked out to resolve the growing disagreement between the developing countries and the CGIAR donor countries. As noted briefly, the 1983 FAO convention voted overwhelmingly to establish an Intergovernmental Committee to monitor and/or oversee the activities of the FAO and the IBPGR, but the powers and responsibilities of the IGC were not specified and membership was open to all. The IGC thus had undefined powers to intervene in the work of the IBPGR and to provide a continuing forum for discussion and, presumably, action. The problem was that the type, timing and effect of the action were unknowns, which considerably raised the level of uncertainty. For example, what kind of understandings would be available for governments funding IBPGR's work? Would newly discovered genetic resources still be available to countries that could not reciprocate with the products of their privately funded breeding programs? Would the IBPGR be able to continue its assignment of preserving genetic resources or would it become another politically active but functionally weak U.N. agency? As the debate has proceeded since 1983, the work of the IBPGR has continued without serious disturbance. By the nature of their mandates, IBPGR and FAO must continue to work together, yet they also need to be somewhat autonomous. An accommodation now seems to be developing, and it appears that a sensible compromise may be within reach.

The agricultural centers under the CGIAR have not been affected by the IBPGR-FAO controversy, but several of the same criticisms have been made concerning them—dominance by donor countries, failure in truly meeting developing country needs, and too close an association with private seed companies. The validity of these arguments is extremely weak. There are nine agricultural centers, all located in developing countries and all working on technical problems with specific application to agriculture in the Third World. The resources being put into their efforts exceed $200 million and would not have rapidly grown in the 1970s without the creation of the CGIAR. Yet, it is argued by critics, the system has somehow "gone wrong" because "companies can take Center material and exploit it for their own commercial purposes—and on a global scale."[9] This is a small part of the story at best. The genetic material developed by the centers is also available to Third World governments in the hope that it will be widely exploited and ultimately raise agricultural output. That is the reason for the research—to exploit the results. Moreover, the centers help developing country governments set up research programs and assist them in exploiting the new varieties. The centers do not assist private companies in their genetic research programs. Quite the opposite occurs; some of the more enlightened companies help the centers. Pioneer Hi-Bred, for example, assists CIMMYT, the corn and wheat center, by multiplying out exotic corn varieties that need to be replenished to maintain the center's gene bank. The center is thus able to devote more of its scarce resources to development of new improved varieties rather than maintenance of existing exotic lines. However, even this is viewed as somehow sinister. In their extreme form, many of the arguments against the existing system rest on the premise that private enterprise is innately evil and that whatever private companies gain in a relationship is a loss for someone else, usually the developing countries.

Such arguments are naive in their conceptualization and incorrect in their facts. The only possible loser in a system that promotes widespread dissemination of seeds is the potential monopolist; everyone else gains. There is no doubt that both the developed and developing worlds have gained enormously from the existing system, with the largest share of benefits going to the Third World. That developed countries, and specifically private firms, have obtained some benefit does not mean that developing countries have received less.

A second factor to consider is the issue of funding. Public budgets everywhere are tight and expenditures are being cut by governments in the North and in the South. In these times, it would be wise to look for allies in defending the present level of funding for the system of international research centers. That firms and individuals in the developing countries benefit indirectly from the programs makes them more likely to support funding of the centers in the years ahead. At this time, attempts to reduce the benefits going to individuals and firms in the North could have disproportionate effects on the political support for continued funding. If one is realistic about helping Third World agriculture, the spin-off benefits going to the countries funding the CGIAR should be viewed as a plus and be encouraged.

Finally, it is worth examining some of the specific facts in the matter of private firms benefiting from the research of the International Agricultural Research Centers. On one level, everyone eventually benefits, in a general way, from everyone else's research. All varieties protected or not can be incorporated into breeding programs immediately upon introduction. Basic breeding pools also are treated by all as public property. Personnel change firms and institutes periodically, and one way or another information is spread. On a more specific level, the facts seem to indicate that there is less mutuality of research strategies between the private companies and the centers than might be expected. In corn, where there is certainly the greatest potential for overlap, the companies and CIMMYT are following different strategies with regard to the developing countries. Given the problems that exist in Third World agriculture and the added cost of hybrid seed, the center is emphasizing improved, open-pollinated corn varieties as the best near-term approach to improving yields through genetic research. The companies are more optimistic and are emphasizing research to develop hybrids that are well adapted and high yielding in Third World settings. If the companies are right, farmers could experience significant yield gains as the new hybrids are introduced. If the companies are wrong, then developing country farmers can fall back on open-pollinated varieties with lower yields and lower cost. The developing countries are thus the beneficiaries of complementary research programs financed by others. While the open-pollinated varieties being developed by CIMMYT may be of some interest to the companies, the pursuit of different strategies makes it less likely that a center-developed open-pollinated variety will be directly useful in the companies' research into hybrids. In the case of corn the beneficiaries seem certain to be the developing countries.

It would be a tragedy if the dispute concerning the IBPGR spilled over into the work of the research centers. As pointed out here, the present system is stacked in favor of the developing countries and has made an enormous contribution to agriculture in the Third World. The benefits that accrue to the developed countries and to private companies are a convenient by-product making the program somewhat more defensible in the donor countries. A compromise or mutually acceptable resolution is needed on the IBPGR issue so that its important work is not disrupted. The rest of the CGIAR system should be recognized as the success it is and be allowed to evolve slowly as new needs and realities emerge.

Developing countries need to become more realistic in their dealings with international seed companies. There are tremendous mutual benefits to be gained. A diversity of genetic resources and seed suppliers is the key to security of supply, high quality seed and competitive prices in any seed market, and these are certainly the objectives of seed policy in the developing countries. A long-term relationship with the leading international seed firms is part of the means to achieve these goals. Another potential gain, recognized by agriculturalists in many developing countries, is the stimulus to competition and the example of experienced efficiency in breeding and seed production that the international firms could give to indigenous seed firms.

PUBLIC RESEARCH PROGRAMS

Within all developed and most developing countries, there are domestic programs of publicly funded agricultural research. At early levels of agricultural development, these are typically the only significant forms of agricultural research. In the United States before the 1920s, commercial plant breeding was largely confined to the home garden market, while agricultural seed was the responsibility of federal and state breeding programs.[10] Only after the introduction of hybrid corn did a private seed industry begin to grow. Even today private varieties dominate a minority of crops in the United States. In most developing countries, the greatest part of seed research is done by the public sector, and these programs are beginning to make an important contribution, often based on initial work done by the international research centers.

This section is concerned with the public sector research activities conducted by governments of the developed countries. Many of these public agencies have long and distinguished records, but are in a difficult transition period. They are still fundamentally important to agricultural progress and, as noted in Chapter 2, calculations indicate that their work has yielded a very high rate of return,[11] yet they have not been able to maintain their funding levels in real terms. The United States is used here to represent the general types of problems confronting public research programs in developed countries. Each country's situation varies slightly, but the general trends are evident. One point to be noted is that government-run agricultural research institutes in Europe have been criticized for the amount of resources they put into flowers and minor crops. European seed producers believe their public research programs are not giving adequate attention to the basic crops, and instead are spreading their limited resources too widely to include some peripheral species.

In the United States, and generally in Europe, neither public nor private sector research organizations feel they are in competition with one another. Although some areas of overlap and competition exist, these are the exception. Private breeding programs have tended to concentrate on finished varieties for major crops with large potential markets for the new varieties produced; examples are corn, sorghum, soybeans, sugar beets, alfalfa, and cotton. Public plant breeding attempts to complement and support this thrust. There are four basic rationales for public research programs in agricultural seed, some of which have

been discussed in more detail in the previous chapter:
   (1) the inability of private firms to capture an adequate share
   of the gains of their research and consequently a chronic tendency to underinvest in agricultural research;
   (2) a strong complementarity between education and research;
   (3) a desire to increase competition through the wide disper-
   sion of the benefits of important findings to many companies;[12]
   (4) the need to support the national private industry vis à vis
   international competition and thereby have a secure supply of
   high quality seed for domestic agriculture.

The last rationale is very common in countries where there is a strong
aversion to dependency on a few international suppliers of seed.

Based on these broad reasons, public sector research has tended
to focus on the location and development of new germplasm that will
improve the yields and other performance characteristics of existing
varieties. In the less advanced markets, public sector research is usually
closely aligned with seed production and marketing operations. In the
more advanced countries, these are most commonly performed by the
private sector; public sector research tends to specialize instead in
basic research and other complementary activities, including:
   (1) the development of finished varieties for minor crops and
   limited crop areas;
   (2) basic research into pest and stress resistance properties;
   (3) basic research into nutritional properties;
   (4) basic research into fundamental cellular and genetic properties;
   (5) the development of new and improved breeding varieties
   for broad use in breeding programs;
   (6) the education and training of future plant breeders;
   (7) to function as a source and clearinghouse for publicly available breeding knowledge.[13]

This thrust of public research is widely accepted, but problems
have arisen in recent years because of a lack of sufficient funds to
support adequately public research programs. The perspective of
breeders is that federal and state funding for public seed research has
not kept pace with inflation for almost a decade, and therefore real
resources available have declined.[14]

In considering total public agricultural research in the United
States, state-run activities are approximately twice the size of the federal programs—$746 million compared with $370 million in 1976.[15]
This includes aspects of agriculture that go well beyond plant breeding
research, e.g., harvest technology, mechanization, pest and insect control. Federal research is funded entirely by the federal government.

Funding for almost 60 percent of the state programs comes from state governments; about 35 percent comes from the federal government; and 6–7 percent is derived from private industry and foundation sources. Public plant breeding research in the United States is conducted by federal government agencies under the U.S. Department of Agriculture and the many state universities and state agricultural experiment stations associated with universities. Although precise data on the level of U.S. public and private seed research is lacking, various estimates are available. High side estimates indicate that, in 1980, private sector research on seed was about $150 million compared with public research expenditures of about $250 million, for a total of roughly $400 million.[16] Other estimates put private plant breeding research in a range of $60 million to $155 million in 1979.[17] It would appear that, in the early 1980s, the preponderance of plant breeding research was still done by the public sector in the United States.

For most crops, the majority of the varieties in use have been developed by public research. Even in corn, with over 80 percent of commercial production based on private varieties, "inbred lines developed by public breeders continue to account for more than half the hybrid . . . seed production in the United States."[18] Therefore, the public research effort is of great consequence in all crops, even when it does not generate the finished varieties. The listing of the apparent needs that future public research ought to address is long and challenging, and there is no indication that any of the rationales for public sector agricultural research have diminished. What has diminished is the real level of funding from traditional sources, even as the need for more basic research is building. Plant breeders are strongly united on this issue: "The 'well' of basic knowledge is running dry and unless we begin to replenish it soon, the productivity of future applied research will decline."[19]

As a consequence of fiscal stringency, public research programs have been under pressure to cut expenses and to recover costs through user fees or other formulas. Several alternatives have been explored, but all present problems in terms of at least potential conflicts with one or more of the basic rationales for public sector research. One possibility is for public breeders to release more finished varieties and for seed companies to market them under a royalty arrangement. This would help raise revenue, but has some obvious drawbacks. It puts public research into more of a competitive position vis à vis private research. It leads public research into the uncharted waters of pricing products, negotiating contracts, collecting fees, and hiring administrators and staff to handle those functions, and it does nothing to help replenish the depleted "well of basic knowledge." It may compromise cooperation with private seed companies. It could create conflicts of interest between the basic rationales for public agricultural research—

especially enhanced competition—and the desire for expanded royalty revenues. Despite these problems, some public research agencies,
particularly state agencies, are restricting release of more new varieties by requiring some form of royalty payment.

Another possibility being considered is to undertake privately contracted research where the sponsor retains the commercial rights to
the products of the research. This practice would also violate the basic
premises under which public agricultural research has developed, and
it introduces even more difficult pricing, contracting and administrative
problems. Measures of this kind are stopgap at best. They lead to dis-
continuities in research programs, do not address long-term basic research needs, and may result in ethical problems. Still another possibility
is an "industry affiliates" program similar to practices that have developed in the electronics industry. Several firms contribute to an agreed-
upon ongoing research program and participate in periodic joint sem-
inars to discuss results that are eventually published. It is doubtful
that this could be a large-scale factor in filling the financing gap, but
it does conform to the traditional role and relationship of public research. Only time will tell where the solutions lie, but it is clear that
mounting financial constraints are pushing public agricultural research
agencies in new and probably undesirable directions.

The basic research needs are clearer than ever, but the political
will is not there. With the advent of genetic engineering, there may
be a permanent shift of the "cutting edge" of basic agricultural research away from the public sector and the universities. Scientists who
are having to spend a large share of their productive time coping with
funding problems will, in time, desert the public sector for private lab-
oratories and research stations. The ultimate consequences will be
failure to maintain acceptable levels of competition, efficiency and
yield gains in the seed industry. With a long list of basic needs in conventional plant research and whole new fields opening up in biotechnology and genetic engineering, this is not the time to be cutting real
public sector investment in genetic research. Documentation is overwhelming that the annual social rate of return to genetic research is
in excess of 30 percent. Total agricultural research is presently only
2 percent of federal research spending, yet even with the enormous
opportunities presented by the new biotechnologies, real funding levels
have been falling for nearly a decade. Unless this trend can be reversed
soon, the effectiveness of public sector agricultural research will begin
to deteriorate noticeably. And once the deterioration is established,
it will be doubly difficult to turn around. To a greater or lesser degree,
most other countries' public research programs in plant genetics are
wrestling with similar problems. Events in the United States are particularly important because of the high level of public research ($250
million versus $175 million for all of the international research centers)

and because of spillover effects via international trade and international flows of knowledge. Similar financial constraints are evident around the world.

PRIVATE SEED COMPANIES

Despite growing interest in the seed industry and widespread anticipation that it will be increasingly important to world agriculture in the years ahead, there is a surprising lack of systematic data or analysis on seed and the seed industry. In the United States, the Department of Commerce does not consider it an industry as its product is not manufactured, so it does not gather statistics on the subject. The Department of Agriculture tends to concentrate on other types of farm inputs and has reduced its data collection on seed. In 1981, the Federal Trade Commission said that "an extensive literature search and inquiries with a dozen agricultural economists . . . specializing in farm inputs have turned up no comprehensive industry studies."[20] Two monographs written in the early 1980s did attempt to compile data, which undeniably have helped to fill the information void;[21] however, some analysts of the industry have argued that serious methodological problems and data errors reduce the analytic value of these works.[22] There is no question that a major information gap exists regarding the past, the present and the prospects of this industry, both in the United States and worldwide, and the need is great for balanced analyses of the important issues facing private seed companies. Experienced seed breeders know the history and understand the current situation, but little has been written and published, and this lack of available public information on seed will undoubtedly persist for some time. This section is an attempt to collect the available information, to familiarize the reader with the industry, and to begin to sort the evidence and analyze the issues.

The emergence of the private agricultural seed industry has been a phenomenon of the last 50 years in the United States and Europe. Entry into its modern phase of dominance by integrated, well capitalized, professionally managed firms is much more recent. The modern era of corn is several decades old in the United States, but has yet to occur in many other crops. Conditions also vary by country. In Western Europe private seed companies dominated markets for the key crops in almost all countries by the 1960s. Private companies formed by, or based on, agricultural cooperatives are very much a factor in Europe, particularly in France, and have no real counterparts in the United States. However, these firms are small (only six had 5 percent or more of the total European market in 1970) and are highly dependent on their home market and/or a single crop. Concentration in individual crops is sometimes moderately high, but in 1970 the biggest firm in Europe had only 7–8 percent of total seed sales. In the 1960s, the U.S. corn seed market was dominated by private firms and had stabilized in the sense that the biggest companies in the 1980s were generally the leaders then. Although ranks have changed, with most absorbed into large nonseed firms in the 1970s, the same companies and brand names are still the most visible. For other U.S. crops, public varieties dominated in the 1960s, and this is still true in a large number of cases. In 1980, for example, privately bred varieties of soybeans were planted on a minority of total acreage; major public breeders developed over 70 percent of soybean seed.[23] In the 1960s, the private seed industry was dominated numerically by many small, family-owned, regional, or specialized firms. Incomplete data support the idea that it was a highly fragmented industry with extremely low concentration levels. Aside from the U.S. corn seed companies, the overwhelming preponderance of research was done in the public sector, and private seed companies' operations were usually limited to production, conditioning, distributing, and marketing. During the 1960s, fewer than 50 U.S. seed firms employed plant breeders, and if the three largest companies are discounted, the average firm employed only two or three breeders.[24] Inclusion of the three largest firms raises the average significantly, so there is little doubt that most of the private sector research was done by a very few firms. No comparable data exist for European firms, but their use of plant breeders was probably similar to that of U.S. companies.

As noted above, until 1970 the seed industries in Western Europe and North America were characterized by public organizations conducting all of the basic and most of the applied research in plant breeding; a small number of large integrated seed firms emphasizing applied research, production and marketing; and many small firms doing only production and marketing operations. These small, generally family-run firms used breeding varieties developed by the public sector and concentrated on one or two crops in a particular region. The division is somewhat explained by the seeds and their characteristics. The larger private companies have been most successful with hybrid corn and sorghum, while public sector research and many small producing and distributing firms have tended to dominate open-pollinated markets.

As described in the previous chapter, research is a long and costly process requiring 10 to 15 years of well directed effort even to have a chance to pay off. Seed production is not as long term, but it requires some skill and quality control to ensure that the seed harvested, conditioned and sold is genetically correct, as labeled and represented to the buyer. Producing the seed is generally done by contracting farmers who are paid a multiple of the value of the basic commodity price. In Europe, this procedure is an important current source of the poor competitive position of seed producers in the European Economic

Community. The prices of many basic commodities, including corn, are subsidized by the Common Agricultural Policy, while seed prices are not. Therefore, Western European seed producers must pay the subsidized cost plus the markup and still price their seed competitively vis à vis producers outside the EC who do not have to pay subsidized prices to their contracting farmers. After the seed has been produced by the contracting farmer, it is cleaned, conditioned and bagged for distribution. During production, quality control must be maintained to meet company and certification standards that include several inspections, determination of germination rates and labeling requirements. It is at the production and conditioning level that the number of seed firms increases dramatically. Once the research hurdle is removed, the barriers to entering the seed trade are very low.[25] When publicly developed varieties are available, all growers who meet basic standards can produce, sell and compete. Moreover, the certifying and labeling system guarantees standards to farmers and minimizes the opportunities for firms to raise margins. When the finished variety has been publicly bred, there is little justification or opportunity for seed growers and traders to ask high prices from the farmers.

Marketing seed is based partly on confidence and familiarity with the supplier and partly on the performance of the variety relative to the alternatives. Performance is a complicated function that boils down to perceived profitability. This depends largely on harvested yield, but also on the quality of the harvest in terms of what the market pays for, i.e., size, appearance, lack of damage, storability, and moisture content. Every crop has its own set of criteria. The economic concept at work here is that of value added. Seed growers can theoretically charge an amount up to, but not exceeding, the value they have added to their product relative to the alternative. The ultimate alternative is "bin run" or farmer-produced seed saved from the previous year's production for open-pollinated crops and is normally 20 percent more expensive than the commodity price of the grain held back and used for seed. This sets a benchmark as to what can be charged commercially for open-pollinated seed. Hybrids have a yield boost of 20–25 percent over saved seed due to greater hybrid vigor. Therefore, hybrid seed is worth considerably more, i.e., up to as much as 20 percent of the total value of the final grain output. Open-pollinated varieties do not have this kind of superiority over farmer-produced seed, and therefore they cost comparably less. Hybrids generally bring 10 to 20 times the value of the commodity depending on the yield characteristics and complexity of the cross.[26] By comparison, open-pollinated seed, such as wheat and soybeans, sell for only about 2.5 times the commodity price, which is principally compensation for value added in cleaning and conditioning the seed.

For North American and Western European farmers, seed is a

small part of the total cost of farming, and demand for seed is not particularly price sensitive. Calculations for hybrid corn in the Midwest United States show seed costs to have fallen from about 4 percent of total costs in 1970 to 3.5 percent in 1980. In absolute terms, seed costs rose during the period, but not as fast as other variable costs (fertilizer, pesticides, fuel, and machinery repairs), which totaled about 30 percent. Fixed nonland costs (nonland interest, machinery depreciation and so forth) were 35 percent, and land costs were about 32 percent.[27] With the recent difficulties in U.S. agriculture, these proportions may have changed somewhat, but seed is still a small part of the total farming costs. For open-pollinated seed, the relative cost would be somewhat less, but because value added is less, farmers tend to buy a smaller proportion of their seed needs from the industry. Wheat farmers in North America typically supply 90 percent of their seed needs from the previous year's crop.

In developing countries, the situation is somewhat different. Many farmers are operating at or near subsistence levels and do not have the resources or skills to utilize improved seed. Although better seed is integral to the overall strategy of improving Third World agriculture, there is commonly an initial reluctance on the part of the farmer to pay market prices for improved seed because historically seed has been a free input. There is also the factor of risk which increases in two ways for the farmer: first, the new varieties may not be as well suited to the farmer's fields as the tried-and-true seed that has sufficed in the past; second, to achieve high returns on the investment in better seed, the farmer should fertilize, use pesticides and/or have access to irrigation, and these all require money, knowledge and effort. The first is an agricultural risk, the second an economic one. In both cases, the farmer is changing the approach used, with the possibility of winning big—or losing big.

The concept of improving seeds, that is, adding value to them, is new and not readily accepted in traditional agricultural settings. This is a second important barrier that seed agencies and companies have to overcome if they are to sell to a broad range of farmers in developing countries. Until 1980, improved seed had principally affected a minority of the more prosperous and more able farmers in the Third World, but this impact is now slowly widening as high yielding rice and wheat varieties in particular are being accepted. The great challenge for the seed industry is to bring improved seed to the mainstream of Third World farmers in the 1980s and 1990s. As noted in the previous section, the first step in this process is a realization by governments in the developing world that seed companies can make an enormous contribution to the improvement of agriculture and should be fitted into their overall program of agricultural development. For their part, firms must be ready to adapt to the countries' needs and fit into a larger development strategy.

Size and Nature of the Seed Industry

Estimates have been made for the total seed market in the United States in 1982, and several important conclusions emerge from this data, presented in Table 3-2. Multiplying the first and third columns gives an estimate of the total market value of the seed used in the 25 categories of species (column 4). This came to roughly $4.65 billion in 1982, which is very low compared with almost any other industry. Moreover, this figure included the value of seed retained by farmers and therefore exceeds sales by the industry. An adjustment for this is made in columns 5 and 6, which demonstrate the proportion of the market supplied by the seed industry. This estimate shows fully 90 percent of wheat seed needs being met by farmers out of the previous year's crop. Similarly, farmers supplied themselves with 45 percent of their soybean seed, 50 percent of cotton and barley seed, and 60 percent of oat seed. The seed industry supplied $3.06 billion or roughly two-thirds of the value of seed used in the United States in 1982. This figure also needs to be adjusted to determine the contribution of the seed industry to agriculture. Specifically, the commodity value of the seed produced must be deducted to identify the genetic value added by the seed industry. Columns 7 and 8 adjust for the share of the value of seed sold, which is attributable to the commodity value of corn or wheat. The contribution of the seed industry is the genetic enhancement it achieves over and above the basic value of the crop as animal feed or human food products. When the seed is not used as a grain or consumed as such (e.g., flowers, beets, vegetables, lawn seed), then the value added is the full amount of the stock of seed supplied by the seed industry. After deducting the commodity value of the seed, the total value added by the seed industry is $2.38 billion.

Although it would be useful to separate out private and public contributions to value added, this is not possible. Most public varieties are distributed by private firms that provide conditioning and marketing services. The research contributed by the public sector cannot be separated out. In a perfectly competitive market, the price of seed would be driven down to the point where the farmer would not pay for public research at all. In this case, the total of $2.38 billion represents only private value added; public value added is a free good to farmers and consumers. In an imperfectly competitive market, or one where royalties are paid to public breeding organizations, then some of the value added shown is attributable to the public sector, but this cannot be estimated.

The first important point to note is that value added by the seed industry is only slightly more than half the total value of seed used. Some 34 percent of seed value is supplied by farmers out of seed saved from previous crops, and another 15 percent must be counted as the

TABLE 3-2. ESTIMATE OF MARKET FOR AGRICULTURAL SEED STOCKS IN U.S. AND VALUE ADDED BY GENETIC SUPPLY INDUSTRY (G.S.I.), 1982

Seed Stock	Amount Used	Unit	Average Price Paid per Unit $	Total Market Value	% of Market Supplied by G.S.I.	Value of Stock Supplied by G.S.I.	% of Value Added by G.S.I.	Total Value Added by G.S.I.
Corn seed	23,534	80,000 kernels	$ 55.00	$1,294,370	95	$1,229,652	95	$1,165,169
Grain sorghum	1,996	50 lbs.	24.50	48,902	95	46,457	90	41,723
Forage sorghum & crosses	540	50 lbs.	16.00	8,640	100	8,640	NA	
Soybeans	68,120	bu	14.00	953,680	55	524,524	50	262,262
Wheat	110,990	bu	8.00	887,920	10	88,792	50	44,396
Cotton	7,102	50 lbs.	20.30	144,171	50	72,085	90	64,539
Sunflowers	405	cwt	180.00	72,900	95	69,255	93	64,253
Subtotal				3,410,583		2,039,405		1,642,342
Alfalfa	75,000	lbs.	1.85	138,750	97	134,588	35	
Subtotal				3,549,333		2,173,993		
Oats	37,449	bu	4.40	164,754	40	65,901	55	35,946
Barley	17,084	bu	6.00	102,504	50	51,252	50	25,626
Rye	2,559	bu	6.50	16,634	80	13,307	38	5,118
Flaxseed	680	bu	11.00	7,480	90	6,732	36	2,448
Rice	4,802	cwt	25.00	120,050	85	102,043	46	46,940
Peanuts	1,094	cwt	70.00	76,588	70	53,606	64	34,461
Sugar beets	2,763	lbs.	5.00	13,816	100	13,816	100	13,816
Sweet clover	9,000	lbs.	.25	2,250	95	2,138	100	2,138
Red clover	28,500	lbs.	1.25	35,625	95	33,844	100	33,844
Timothy	10,000	lbs.	.90	9,000	95	8,550	100	8,550

TABLE 3-2 Continued

Seed Stock	Amount Used	Unit	Average Price Paid per Unit $	Total Market Value	% of Market Supplied by G.S.I.	Value of Stock Supplied by G.S.I.	% of Value Added by G.S.I.	Total Value Added by G.S.I.
Ryegrass	230,000	lbs.	.30	69,000	95	65,550	100	65,550
Tobacco	135.11	oz.	28.00	3,783	90	3,404	NA	—
Packet seeds	—	—	—	20,000	100	20,000	100	20,000
Bedding plant seeds	—	—	—	7,000	100	7,000	100	7,000
Cut flower seeds	—	—	—	1,000	100	1,000	100	1,000
Lawn seeds	350,000	lbs.	1.00	350,000	100	350,000	100	350,000
Vegetables				100,000	85	85,000	100	85,000
Total				$4,648,817		$3,057,136		$2,379,779

NA = not applicable.

Source: Edwin M. Kania, Jr., *Pioneer Overseas Corporation Case Study* (Cambridge: Harvard Business School, Case 4-583-070, 1982).

commodity value of the seed sold. Therefore, even in the United States, markets remain to be served if the industry could develop the right seed and benefit from its contribution to value added by retaining a larger share of the annual seed needs of the total market. Wheat and soybeans seem to be markets that have potential, but may require hybrids to achieve high annual sales. Hybrids would preclude rapid reversion to farmer-supplied seed, and the seed industry would benefit from the genetic improvement it financed and developed. A serious effort is being made by several companies to improve soybean yields and expand their share of the market. This will be discussed in more detail below, but the results are worth watching, to see if the new varieties are significant improvements and if the companies can benefit from the value that they manage to add to soybean seed.

The second point to note is the concentration of the seed market in a few items, one of them not even an agricultural crop. The share of the total market for each of the leading categories can be found in Table 3–3, which shows lawn seed emerging as a large market for the seed industry. Corn leads all other categories, and its dominance increases as adjustments are made for retained seed and the commodity value. Corn's domination rises from 27.8 percent of the total market to almost half of value added by the seed industry. Soybeans and wheat are large open-pollinated crops, but the seed industry has not been able to make the same kind of proportionate contribution to value added as it has in hybrid corn. Lawn seed is not entirely comparable with the other big markets because it is not a harvested commodity. If lawn seed and the other noncommodity crops are deducted from value added, the total drops from $2.38 billion to $1.79 billion, and corn's share moves up to 65 percent of value added in the seed industry.

TABLE 3–3. PERCENTAGE SHARE OF U.S. SEED MARKET BY CROP, 1982

Crop	Total Market Value	Value Supplied by Seed Industry	Value Added by Seed Industry
Corn	27.8	40.2	48.9
Soybeans	20.5	17.2	11.0
Wheat	19.1	2.9	1.9
Lawn seed	7.5	11.4	14.7
Oats	3.5	2.2	1.5
Cotton	3.1	2.4	2.7
Rice	2.6	3.3	2.0
Sunflowers	1.6	2.3	2.7

Source: Kania, *Pioneer Case Study.*

In terms of making private markets work and focusing resources on agricultural research, the hybrid characteristic of corn has been the most outstanding success of the seed industry. Corn seed breeders were able to retain a high proportion of annual sales in the market and plow profits back into research. Over time, the research added value to the product and margins rose, enabling more resources to go to research. This is an excellent case of how the patent system ought to work. An inventor is rewarded and proceeds to invent even better products using the gains earned from the first invention. Moreover, with the built-in patent protection resulting from hybrid vigor, there is little or no chance of evasion of the patent laws or enforcement of expense. Another attraction is that research can be financed directly out of its own earnings and does not have to contend with the vagaries of the political budgetary process. The problem with open-pollinated crops—even those with plant variety protection—is that the seed firm that makes a breakthrough may not benefit appropriately. It can produce a variety with significantly improved performance yet end up losing most of the value added because farmers buy the improved seed once from the breeder and use retained seed thereafter; the farmer and consumer benefit, but not the plant breeder. The result then would be to curtail private sector breeding efforts in open-pollinated crops and reduce their rate of yield improvements, as compared, for example, with corn. This, in fact, seems to be the case as corn yields in the United States roughly doubled between 1960 and the early 1980s, while wheat yields rose only about 60 percent.[28] Private research surveys by crop, discussed in some detail later in this chapter, indicate that corn breeding receives between 40 and 50 percent of the seed industry's total efforts in traditional plant breeding.

For Western Europe, Japan and Australia, a schematic of the genetic supply industry would look roughly like that of the United States. Due to climate variations, the importance of crops would vary and the share of seed retained by farmers would be somewhat higher, but the other patterns would be present. Hybrids would have high value added by the seed industry and a high share of their markets, while open-pollinated seed would not. In developing countries, the experience is much more varied, with most farmers retaining unimproved or only slightly enhanced seed, rather than buying improved agricultural seed on the market.

There is an enormous potential market for improved seed. For corn, only 23 percent of worldwide acreage is in the United States; therefore, a simple projection shows a potential market roughly three times the U.S. market. Of this, the largest acreage allocations to corn are to be found in mainland China (15 percent), Brazil (9 percent) and South Africa (5 percent).[29] Progress in bringing improved seed to developing countries has been slow, as discussed. The high yielding varieties developed by the International Agricultural Research Centers were made available to developing countries at little or no cost for basic research. Help has also been provided in applied research and local production and distribution. Yet, it still took more than a decade for the new seed to be accepted widely by the more able farmers. Private companies face the same obstacles in terms of technical problems and the slowness of the farmer to change, but they also have to contend with the ambivalence, or even hostility, of the government. For these reasons, progress will be slow, and both the governments and the companies will have to adjust their approaches if improved seed is to realize its potential in the developing world over the next two decades.

Looking back on the evolution of the private seed industry, the development of hybrid corn seed in the mid-1920s inaugurated a new era. To date, that milestone has principally affected the corn and sorghum seed trade, where successful hybrids have been developed; it has hardly touched wheat or many of the other crops listed in Table 3–2. Corn clearly stands out in the seed industry. Corn has entered the modern age in that it is able to fund much of its own research and initiate a positive feedback sequence in which everyone benefits— breeders, farmers, consumers—and all continue to play their roles in the sequence. The farmers benefit from higher incomes, the consumers have more and better food, and the breeders receive fair returns for the knowledge they developed.

The Changing Structure of the Seed Industry

In the 1960s and early 1970s, several events initiated a period of transition in the private seed industry. Many feel it is a transition to a new era in which the success of corn can be approximated, if not replicated, in other crops. Others view the transition with alarm and feel that the private seed industry is changing for the worse with unfortunate consequences for world agriculture. The events that have changed perceptions of the seed industry over the past 15 years will be reviewed here and the results examined to determine the possibilities that lie ahead. Some of the perception-changing events have been noted, others have not yet been discussed. Their net result was to make seed companies more attractive as business options and to attract outside corporate investors. The key events that brought about the changes in perception are listed, more in chronological order than in terms of importance:

(1) the success of the Green Revolution and the realization that genetic research could have enormous worldwide implications;
(2) the agricultural commodity price boom that began in the late 1960s and persisted through most of the 1970s;

(3) legislative changes in many developed countries in the form of plant variety protection aimed at giving breeders a better chance of benefiting from their research investment;

(4) technical discoveries focusing on cellular properties that offered the promise for dramatic breakthroughs in biogenetic engineering;

(5) cutbacks in public plant breeding research programs coupled with the expansion and continued success of the private sector;

(6) energy price rises in 1973 and 1979, and an end to the energy era in world agriculture;

(7) the desire for large, successful industrial firms to diversify into new sectors with high growth potential;

(8) growing environmental and regulatory concerns and costs, and the desire to move into relatively clean, unregulated types of business activity;

(9) declining profits and growth potential in many of the traditional industrial sectors;

(10) increased worldwide concern regarding the reduction of genetic diversity and the greater awareness of the value of existing germplasm.

In the late 1960s, some larger industrial firms were beginning to develop an interest in seed companies. Seed companies offered great potential for diversification, growth and high returns. Those firms showing the most interest tended to be in pharmaceuticals, chemicals, petroleum, and food processing. During the 1950s and early 1960s, agricultural prices had been stable or declining, and food-related industries were not viewed as attractive investments. The Green Revolution had shown what could be done, and within the seed industry some firms had increased research budgets, but outside interest was just beginning to be evident in the late 1960s. Then, in the early 1970s, events occurred that made investment in the genetic supply industry more and more attractive.

From the point of view of the seed companies, merger with a larger company that could supply resources for research often made good sense. Even the large seed companies were small compared to the challenge of bioengineering, and they needed capital to have a realistic chance of competing in the new environment. Smaller seed companies were facing even greater problems with the public research cutbacks. As the older family members withdrew from these small family-operated firms, the next generation often preferred to cash in on the company's existing value rather than face increasingly stiff competition in the coming years. From the perspective of the firm taking over, the acquisition also made sense. In addition to the 10 perception-changing events noted above, a variety of specific motives came into play. Firms with existing research strengths saw the potential for cross-fertilization between the plant breeders' genetic work and ongoing research in pharmaceuticals or chemicals. By acquiring several seed firms specializing in different crops or regions, economies could be achieved in production and marketing. There was also the potential for complementarities in the marketing of seed and other agricultural inputs, notably chemicals. Most analysts, however, believe the primary motivation for takeovers is diversification into a new field with good existing growth and profitability and enormous potential over the next quarter century.[30]

The amount of conglomeration that has taken place is striking. In 1970, almost all seed companies were independent or related to another seed company. By 1982, the large U.S. seed companies, with only a few notable exceptions, were part of larger nonseed corporations. In Europe, the trend is proceeding somewhat more slowly as many existing seed companies are cooperatively owned, making acquisition more difficult and costly. Nonetheless, several of the leading cooperative-owned companies may soon be facing problems and could use an injection of professional management and/or expanded research capabilities.

No comprehensive listing of total mergers and changes in company status exists, but such evidence as has been collected is indicative of the activity that occurred in the 1970s. For the United States, a listing of 100 mergers and acquisitions has been made and, where possible, dated. Of these, only 2 occurred before 1970, 16 between 1970 and 1973, 45 between 1974 and 1979, 19 in the 1980s, and the rest of the dates were not reported.[31] Another attempt to enumerate changes in the status of seed companies in 18 of the more advanced countries reported 762 corporate changes since the late 1960s.[32] The specific details may be obscure, but the facts are that there has been a wave of mergers and acquisitions of seed companies since 1970. Although the wave now appears to have abated, it is probably not exhausted. Depressed equity prices could add to the reasons mentioned above, making it attractive for large firms to invest via the acquisition route rather than to create completely new operations. It is not hard to list the pluses and minuses of the acquisition surge of the past 15 years, but it is very difficult to draw the balance at this point in time. On the plus side, there are the following arguments:

- greater available resources for increased research;
- greater access to information worldwide regarding new varieties and other developments;
- broader facilities and capacity for testing and marketing;
- research cross-fertilization from other disciplines.

On the negative side, there are the following concerns:
- seeds will be treated like any other profit center and be required to show results on a quarter by quarter basis;
- reduced competition may have several harmful consequences;
- management of research activities is always a challenge, and genetic research, which is both art and science, may not respond to existing corporate procedures;
- the corporate hierarchy may not understand the seed business and the bureaucracy may reduce effectiveness.

The record of the conglomerates is mixed to date. In some cases, the acquired seed operations seem to be thriving, but in others, problems are emerging. In a few cases, larger firms are trying to dispose of their seed divisions.[33]

A listing of the largest international seed companies is revealing on several counts. It shows the extent to which nonseed parent firms are now a factor in the seed industry, compared with a very low profile for such outsiders in 1970. It shows that in Europe several sizable firms still remain as specialist seed companies. Most important, it shows the extremely low concentration ratios found in the world seed market. The list is presented in Table 3-4 with some basic data on the companies' origins and operations. Of the large American seed companies in the table, only Pioneer Hi-Bred is still independent; among the European firms, four of the largest are still independent. There are five other firms clearly of the conglomerate type, in which seeds are a small and new part of overall operations, including Shell, Sandoz, Pfizer, Ciba-Geigy, and Upjohn. The remaining four firms are a mixture: Cardo has most of its sales in seed, but absorbed Weibull in the process of its expansion; Cargill has had seed operations for some time, but acquired several additional firms in the 1970s; and seed activities are a large part of the sales of both of the Dutch companies. Further mergers and acquisitions could change this listing, but for now it appears that several of the large European companies may maintain their independence. Pioneer will also probably stay independent.

Regarding concentration, the seed industry remains remarkably fragmented relative to many other industries. Estimates for the market value of the world seed trade are in the range of $45 billion to $50 billion for 1983, with the United States accounting for about 10 percent.[34] Relating this to seed sales by the leading firms, the top five had only $2 billion in sales for a remarkably low ratio of 4 percent. It can be argued that this is the wrong measure, that farmer-retained seed should be taken out as should the commodity value of the sales. Removing the retained seed reduces the value supplied by the seed industry considerably, perhaps by one-half, but the matter of the commodity value cannot be addressed because it is not known. And the

TABLE 3-4. DATA ON 14 LARGE SEED COMPANIES

Company	Nationality	Principal Operations	Total Sales (U.S. $ Mill.)	Seed Sales (U.S. $ Mill.)
Royal Dutch/Shell	UK/Dutch	Petroleum	82,291	650
Pioneer Hi-Bred	U.S.	Seeds		557
Sandoz	Swiss	Pharmaceuticals	2,946	319
Cardo	Swedish	Agro. ind.	440	285
DeKalb/Pfizer	U.S.	Pharmaceuticals and oil and gas		187
Claeys-Luck	French	Seeds		155
Upjohn	U.S.	Pharmaceuticals	1,828	139
Limagrain	French	Seeds		130
Ciba-Geigy	Swiss	Pharmaceuticals	7,061	107
Suiker Unie	Dutch	Agro. ind.	353	100
K.W.S.	German	Seeds		80
Cebeco	Dutch	Agro. ind.		65
Svalof	Swedish	Seeds		55
Cargill	U.S.	Agro. ind.	15,000	50

Source: Adapted from Pat Ray Mooney, "The Law of the Seed," in *Development Dialogue*, Vol. 1-2 (Uppsala, Sweden: Dag Hammarskjold Foundation, 1983). Data appear to be for 1983.

companies' sales figures would have to be reduced, and these are also unknown. Even if market sales are used, excluding restricted markets in developing countries, estimates indicate a value of $12 billion to $15 billion, which gives a concentration ratio of 13 to 17 percent for the top five firms, still remarkably low compared to other industries.

There are three possible explanations. First, the seed industry simply does not have economies of scale, as do most other industries. Economic success, as in construction or baking, is not helped by large-scale operations, thus most firms tend to stay small, with ease of entry or exit, and considerable emphasis on local markets. This certainly characterized the seed industry in past decades, except perhaps for hybrid corn.

A second explanation is that the concentration process is not yet over. There may be more mergers, or there may be a shakeout of weaker firms and concentration by attrition, as happened in the U.S. automobile industry in the 1930s and again in the 1960s. It is unwise to be dogmatic on the subject, but the initial surge of mergers and acquisitions may be largely over, with a shakeout period coming. In fact, several multinational firms including Union Carbide, Olin Matheson, W.R. Grace, Purex, FMC, General Foods, and Central Soya, have divested their seed operations. Some conglomerates will succeed with their seed strategies and some will not. Some firms will achieve biogenetic breakthroughs that are commercial successes, but most will not. This process will determine the look of the genetic supply industry in the years to come, and several familiar faces could be missing by the year 2000.

The third possibility is that the wrong level of concentration is being measured: what is relevant is a specific crop in a specific country. International concentration ratios are always lower than national ratios, and it is at the national level where competition does or does not take place. Because the data have existed for more than a decade and because it is one of the more concentrated sectors of the U.S. seed industry, a closer look at the U.S. corn seed market over the period 1973 to 1983 is informative. Table 3–5 presents market shares for the top eight or so firms. The principal conclusion is that Pioneer has expanded its share of the market by over 50 percent, rising from 23.8 percent in 1973 to 36.9 percent in 1980, with its share rising slowly since then to reach 38.1 percent in 1983. This expansion was at the expense of the four or five next largest companies, whose market shares all fell between 1973 and 1983. The other trend is that, even with Pioneer's surge, the top eight companies lost market share to the smaller companies. Within these general trends, there are several interesting patterns at work, some of which may be related to acquisitions by large nonseed corporations. DeKalb, for example, was even with Pioneer in the early 1970s with about 21 percent of the market, but began

TABLE 3-5. PERCENTAGE SHARE OF U.S. CORN SEED MARKET

Company	Miller Survey*						Pioneer Survey*			
	1973	1975	1977	1978	1979	1980	1980	1981	1982	1983
Pioneer Hi-Bred	23.8	24.6	30.9	26.2	32.9	36.9	34.8	35.0	38.8	38.1
DeKalb[1]	21.0	18.8	15.8	17.9	13.3	13.0	15.9	13.9	12.2**	10.3
Funk[2]	8.8	8.9**	6.4	8.1	6.7	5.7	5.4	5.8	5.2	3.9
Trojan[3]	5.9	6.8**	4.2	5.4	3.8	2.0	—	—	—	—
Northrup-King[4]	6.1	4.7	3.8**	3.3	3.8	4.9	3.4	2.8	2.6	2.5
Cargill[5]/PAG	4.8	3.9	4.1	4.6	3.3	4.7	5.6	5.8	5.4	4.2
Golden Harvest	—	1.8	2.5	3.1	2.9	1.3	3.2	2.8	2.3	2.6
Jacques[6]	—	1.7	1.9	2.1	2.7	2.2**	—	—	—	—
Others[7]	29.6	29.8	30.4	29.3	30.6	29.3	31.7	33.9	33.6	38.4
Largest 8	72.5	70.2	69.6	70.7	69.4	70.7	70.0	67.1	68.3	64.0
Largest 4	59.7	59.1	57.3	55.6	56.7	60.5	59.5	56.9	59.1	54.9

*Due to the sample size of the surveys, these percentages are estimates that may vary plus or minus two percentage points.

**Indicates merger or acquisition.

1. Merged with Pfizer in 1982.
2. Acquired by Ciba-Geigy in 1974.
3. Acquired by Pfizer in 1975.
4. Acquired by Sandoz in 1976.
5. Includes PAG acquired in 1971.
6. Acquired by Agrigenetics in 1980.
7. May include firms not listed individually for certain years.

Sources: Miller Agrivertical Unit (1973–80) and Pioneer Hi-Bred (1980–83).

to slip behind in 1975 and by 1982 had lost about half of its earlier market share. At this point, DeKalb may have decided it needed reinforcements and merged with Pfizer. Trojan was acquired by Pfizer in 1975 and continued to slip for several years. In 1982, its share of 1–2 percent was combined in DeKalb's total. Funk was acquired by Ciba-Geigy in 1974 and has declined since then. Northrup-King was acquired by Sandoz in 1976 and after declining for a year or so rose in the late 1970s. It seems to have fallen again in the 1980s.

The past acquisition record is thus mixed for the top five companies. Pioneer continues to expand its share of the U.S. corn seed market, but more slowly. Funk and Northrup-King may have bottomed out after weaker performances between 1975 and 1983. Trojan has been absorbed, and the big question is whether DeKalb-Pfizer can recoup most of its previous position.

Through the 1970s, overall concentration ratios were moderately high for the top four firms, but increased only slowly for the next four firms. Since 1980, this ratio has fallen, even though Pioneer has added some market share. DeKalb, Northrup-King and Funk continued to slip, and the concentration ratio is thus down compared with the 1970s. Increasing concentration at the top of the U.S. corn market is not the result of powerful conglomerates or monopolies; it has to do with strong performance by a successful old-line seed company. The other companies did not seem to be competing on an equal footing with Pioneer and may have turned to larger parent organizations in hopes of upgrading their research programs and overall performance. In the U.S. corn seed industry, the competition has been quite vigorous for more than three decades and is not likely to diminish any time soon. The driving force leading to concentration is thus not a matter of monopoly, but performance. The next question is how successful the conglomerates will be competing with a strong, traditional seed company, and only time will answer that question.

The tentative conclusion, therefore, is that the surge of acquisitions and the advent of the conglomerates into the seed trade have not yet resulted in excessive concentration at any level. Rather, the problem with the new company owners is how well they can perform in developing better seed from their research programs. Can they produce better performing varieties and show continued yield gains? At this point, hope exists, but there are no grounds for optimism. Some analysts, however, feel that these recent developments have not been good for the seed industry. Butler and Marion, for example, concluded that, "on balance, conglomerates have probably had a negative influence on the seed industry by acquiring many previously independent companies, eliminating actual and potential competitors in some cases, consolidating plant breeding activities and, thereby, reducing diversity, and, in some cases, reducing the viability of the seed companies."[35]

PLANT BREEDERS' RIGHTS

The legal protection of intellectual property extends back for several centuries, but the origins of the present system are dated from the Paris Convention for the Protection of Industrial Property in 1883. From this period, the idea of patent and copyright protection sprang up as a formal international concept and spread throughout Europe and North America. For all of its seeming acceptance, the patent idea is, in practice, a compromise between two contradictory principles, as was discussed in Chapter 2. In terms of static efficiency, goods should be priced at their marginal cost. And since knowledge has a zero marginal cost once it has been produced initially, it should be a free good. In terms of equity and dynamic efficiency, there should be a charge for the use of knowledge so that inventors, authors, programmers, breeders, and others will be encouraged to produce more of it in the future. The patent system is an attempt to balance the contradiction by giving inventors proprietary rights for a period of time, providing they put their invention on file at the national patent office. Others may then buy diagrams and descriptions of the invention, but may not legally reproduce it without the permission of the inventor. The patent thus grants to the inventor a limited legal monopoly to exploit that invention.

Patent laws have come under criticism from various points of view. Patent holders often find that patents give little effective protection and can involve expensive legal wrangles. Corporations, particularly U.S. firms, have of late not even patented new products and processes, claiming that it is more trouble than it is worth. In particular, they argue that there is little or no protection against the patented idea being copied by others once a full description is on public record. On the other side, there is criticism that patents are government-created monopolies and enable the corporations or individuals holding most of the patents to exploit consumers.

Plant breeders' rights in the form of legislated plant variety protection are an attempt to extend patent type coverage to new plant varieties created by breeders. Registration of varieties is based on established criteria of novelty, uniformity and stability. The variety in question has to be new, that is, different from other protected varieties in some identifiable way. It must be relatively uniform in that all the plants of the variety must have the same characteristics. And it must be stable, meaning that the characteristics of the variety do not change or mutate readily. Utility or superiority of performance is not a required characteristic for protection, but is of course useful in marketing the variety. PVP legislation in most countries also specifically allows farmers to reproduce seed for their own use and even for sale, as long as it is not on a "commercial" basis. Plant variety protection is thus much weaker than patent protection because the users, including Third World users, may reproduce the product themselves and thereby capture the embodied knowledge. This is equivalent to being able to buy a new machine once and use it to build as many similar machines as one wants for one's own factory. It is argued that the novelty criterion allows imitators to get protection for varieties that are essentially copies possessing a single marginal difference. This also weakens the effect of PVP. Breeders are very unlikely to be able to earn extended monopoly profits on their protected varieties. With open-pollinated varieties, farmers can always retain part of the previous year's crop and avoid monopoly pricing. This tends to weaken the effect of PVP compared with industrial patents or hybrid seed. Hybrid seed producers retain their breeding varieties and thus are protected by nature against other seed companies buying their seed and co-opting the benefits of their research. In the United States, hybrids do not qualify for plant variety protection, but they do in Europe. With nature on their side, hybrid breeders do not have to worry about farmers reproducing their seed and staying out of the market.

During the 1960s and 1970s, PVP legislation was passed in 17 countries, which thereby became members of the Union for the Protection of New Varieties of Plants. Most were European countries, but Japan, New Zealand, Israel, the United States, and South Africa were also parties to the convention. The United States passed PVP legislation in 1970 and has been issuing variety protection certificates since 1973. The motivations for passing PVP legislation no doubt vary from country to country, but are probably well represented by the following:

- "stimulate private plant breeding research;
- allow agricultural experiment stations to increase needed basic research;
- permit public expenditures for applied plant breeding to be diverted to important areas which industry might not pursue;
- give farmers and gardeners more varietal choice, and higher yielding and better quality varieties;
- make national agricultural products more competitive in world markets;
- provide benefits to consumers of crops and crop-products either through improved quality or greater production;
- foster continued breeding of new varieties by university experiment stations which can license them to seed companies for a share of the proceeds."[36]

These points essentially sum up the public arguments in favor of PVP legislation. However, at its core, the case for PVP legislation is based on the recognition of three fundamental realities:

(1) more resources should flow into agricultural genetic research because of the high social rates of return;
(2) more public funds are unavailable due to heavy budgetary
demands for transfer expenditures and limits on tax increases;
(3) greater private research programs will not have the incentives or the funds required unless they can capture a larger
share of the social rate of return for themselves.

All of the above arguments in favor of PVP come from a recognition
of these three realities. If social rates of return were low, public funds
forthcoming or hybrids available in most crops, PVP would not be
necessary. But because none of the above pertain, PVP legislation is
the best means available to obtain more resources for research on plant
breeding.

A review of the PVP certificates granted in the United States since
1973 indicates a heightened interest in plant breeding by the private
sector in certain crops. The data presented in Table 3–6 list PVP certificates by the year in which application was first made and by the
number of certificates issued from 1971 to December 31, 1983 for six
leading crops. The low levels of certifications shown for 1983 should
not be misunderstood. Most applications made in 1983 and many from
1982 and 1981 are still pending; these are shown in parentheses. Certificates issued in future years will certainly raise the numbers for 1982
and 1983. The figures reveal a significant increase in activity in soybeans with smaller increases for other crops. Alfalfa and cotton may
be up slightly, but certainly not by significant amounts. Beans and peas
are up slightly, with the rise stronger in the case of peas, especially
if the majority of the pending applications are granted. Wheat is also
higher, especially as contrasted with pre-1976. Soybeans, however, are
the crop in which the surge in new varieties is most pronounced, with
a tripling in the certificates granted in the early 1980s, compared with
the early and mid-1970s. Given the lag in developing new varieties,
this seems to be good evidence for the stimulative effect of U.S. PVP
legislation in encouraging breeders to develop new varieties of soybeans and, to a lesser extent, wheat and peas. The interest in soybeans
should not be surprising. As shown in Table 3–3, this was the second
largest market after corn and, in 1982, the majority of the seed was
made up of public and private varieties produced by the seed industry.
Wheat was also a large seed market, ranking third overall, but a much
smaller proportion is supplied by the seed industry.

A breakdown of certificates issued through 1982 for 20 crops is
shown in Table 3–7; soybeans again stand out, with wheat in second
place. These data also differentiate between public and private institutions, with 85 percent of soybean certificates going to private firms.
This contrasts with the minority share of acreage planted in privately

TABLE 3-6. PVP CERTIFICATES ISSUED BY FISCAL YEAR OF APPLICATION

	Alfalfa	Cotton	Garden Bean	Peas	Soybeans	Wheat	Total
1971	0	6	14	4	9	1	34
1972	2	21	6	5	8	11	53
1973	2	7	8	6	13	5	41
1974	1	5	10	8	15	11	50
1975	2	7	8	16	11	7	51
1976	3	6	9	7	13	16	54
1977	0	10	16	16	20	15	77
1978	2	10	3	10	20	11	56
1979	2	8	3	8	35	8	64
1980	5	8	15	19	41	13	101
1981	2 (16)	11	14 (3)	11	52	15	105 (19)
1982	4 (2)	7 (1)	3 (4)	12	36 (6)	14 (1)	76 (14)
1983*	0	7 (1)	0 (8)	1 (21)	3 (30)	0 (19)	11 (79)
Total	25 (18)	113 (2)	109 (15)	123 (21)	276 (36)	127 (20)	773 (112)

*The low number of certificates issued in 1983 reflects action not completed in the final processing of all applications submitted through the period.
Note: Applications in parentheses pending as of December 31, 1983.

Source: Prepared from USDA, *Plant Variety Protection Office Journal.* Figures are as of 12/31/83.

TABLE 3-7. MAJOR SPECIES PROTECTED UNDER PVP RANKED BY NUMBER OF CERTIFICATES ISSUED AND NUMBER OF ORGANIZATIONS OWNING CERTIFICATES OF PROTECTION AND PERCENTAGE HELD BY LEADING HOLDERS (As of December 31, 1982)

Species	Number of Certificates Issued[1]	% of All Certificates Issued	Cumulative %	% of Certificates Issued Under Title V[2]	No. of Firms & Institutions Owning Certificates[3]	% of Certificates to Leading Holders — Number of Leading Holders			
						1	2	3	4
1. Soybeans	241 (37)	22.3	22.3	44	42 (9)	12.4	22.8	31.1	39.4
2. Wheat (all types)	127 (36)	11.7	34.0	87	35 (13)	14.4	22.9	31.4	39.0
3. Peas	113 (0)	10.5	44.5	4	18 (0)	34.5	45.1	53.9	61.9
4. Garden beans (all types)	108 (4)	10.0	54.5	7	23 (4)	24.1	39.8	50.9	60.2
5. Cotton	102 (13)	9.4	63.9	67	25 (5)	20.6	30.4	38.2	45.1
6. Lettuce	44 (0)	4.1	68.0	0	10 (0)	25.0	40.9	54.5	68.1
7. Marigolds	25 (0)	2.3	70.3	0	6 (0)	36.0	64.0	80.0	88.0
8. Alfalfa	25 (8)	2.3	72.6	32	10 (5)	48.0	64.0	72.0	76.0
9. Ryegrass (all types)	23 (1)	2.1	74.6	50	13 (1)	13.0	26.0	34.7	43.4
10. Fescues	22 (7)	2.0	76.6	50	15 (5)	13.6	27.2	36.3	45.4
11. Bluegrass	19 (8)	1.9	78.6	58	12 (1)	10.5	21.0	31.3	42.0
12. Oats	16 (2)	1.5	80.1	94	7 (4)	37.5	56.3	75.1	81.4
13. Barley	14 (0)	1.3	81.4	100	8 (1)	28.6	42.9	57.2	71.5
14. Tobacco	14 (0)	1.3	82.7	93	3 (0)	50.0	87.5	100	—
15. Onions	14 (0)	1.3	84.0	0	4 (0)	78.6	85.7	92.8	100
16. Rice	12 (0)	1.1	85.1	67	6 (0)	33.3	50.0	66.7	83.4

TABLE 3-7 Continued

Species	Number of Certificates Issued[1]	% of All Certificates Issued	Cumulative %	% of Certificates Issued Under Title V[2]	No. of Firms & Institutions Owning Certificates[3]	% of Certificates to Leading Holders Number of Leading Holders			
						1	2	3	4
17. Corn (all types)	12 (0)	1.1	86.2	17	6 (0)	25.0	41.7	58.4	66.7
18. China aster	10 (0)	0.9	87.1	0	1 (0)	100	—	—	—
19. Watermelon	10 (2)	0.9	88.0	10	7 (2)	30.0	50.0	60.0	70.0
20. Peanuts	9 (2)	0.8	88.8	100	7 (1)	22.2	44.4	55.5	66.6
Other crops	121	11.2	100.0						
Total—All Crops	1,081								

1. Number of certificates issued to public institutions in parentheses.
2. To be sold by variety name only as a class of certified seed.
3. Number of public institutions holding certificates in parentheses.

Source: Plant Variety Protection Office, USDA, Beltsville, Md.

bred soybean varieties in 1980. The picture that emerges is one of private seed firms deciding that soybeans and, to a lesser extent, wheat were the crops with a large enough potential market to justify a major push in developing new varieties. These decisions were made in 1970, and the varieties appearing in the early 1980s are the result. Soybeans appear to be the crop most favorable for payback to private sector research because of its size and because farmers already are in the habit of buying seed on a regular basis. To succeed, the private sector has to develop superior varieties, convince farmers of their value, price the seed high enough to show a positive return on research, and keep production costs low enough so that farmers buy on an annual basis. If this strategy is successful, the basic PVP goal of drawing more private resources into plant breeding will have been met for the most attractive product. And as a consequence, private funds will be drawn to other crops. If, however, private firms are unable to show a return on their soybean programs, then PVP may fail in terms of its most basic objective, and a different approach to increasing research resources may have to be formulated.

PVP's Effect on Breeding Research in the United States

There has been a certain amount of argument regarding the effect of PVP to date on the level of research and development expenditures by private plant breeders. One side says that PVP has had no sizable effect on private plant research, while the other contends there has been a positive and significant effect. Those belittling the effect of PVP argue that the surge in new varieties of seed since the middle 1970s can be explained by factors other than PVP. They further argue that, relative to sales of seed, the amount of research being done is not much greater.[37]

Several important improvements have occurred in plant breeding; these include:

(1) increased use of winter breeding nurseries in Florida, Hawaii, the tropics, or the Southern Hemisphere, permitting the multiplication of seed two or three times a year with a significant acceleration in breeding programs, facilitated by improved air cargo service in the 1960s;

(2) better utilization of computer information systems to monitor larger numbers of plant crosses for a greater number of characteristics;

(3) a broader range of germplasm now available through the international research centers and public institutions, and better communications among breeders;

(4) the perceived opportunity of developing hybrids for wheat and other crops and the implications of this for protection of

the firm's research findings and annual sales of seeds[38]; (5) increased mechanization of breeding and testing procedures, such as the use of small plot planters and harvesting equipment.

Taken together, these factors should in time improve the productivity of breeding; that is, more varieties should emerge from a given level of resource expenditure. This should raise the social rate of return, so that it is worthwhile for society to invest more in plant breeding. To argue that because these factors have come into play, PVP is not needed and has not contributed to increased plant breeding is to misunderstand the difference between ends and means and social and private returns. Private firms do not buy computer systems, charter jet cargo carriers, design special plot equipment, or put people to work testing new germplasm sources with the development of new varieties as the end product. Public research may view this as the final goal, but private companies must earn their way, and new improved varieties are a means to achieve this end. Private firms have to be able to sell their newly developed varieties, cover costs and show a return to their investors. Technically the new factors are an important part of a firm's ability to produce more varieties, but there is little point in developing a good new variety if it is not protected from being stolen. While the technical improvements in breeding should raise the social rate of return to plant breeding, more investment is not guaranteed unless the firm receives an adequate part of the total social benefit. Therefore, agreeing that technical improvements, some with significant startup costs, have increased the productivity of plant breeding does not imply agreement that PVP is unimportant. Introducing new techniques and equipment into research or production operations is always a difficult and costly process. It is unrealistic to expect that private companies or individuals will go to great effort and take great risks in developing new plant varieties without the expectation that the legal system will protect the results of their research from theft and other forms of expropriation.

The link between plant breeding research and development and PVP legislation hinges closely on how expenditures on R&D are deflated, i.e., adjusted for inflation. Arguments against PVP are inclined to use seed sales as the deflator. They take the amount firms spent on research and divide by the total value of gross sales (not net income) as an estimate of real R&D effort. The results of this type of calculation are shown in Figure 3–1 for four groups of companies reporting research programs as of 1960, 1967, 1970, and 1976. (The four sets of firms are respondent groups that began their research reporting at different times.) Each group shows a trend upward in research between their first and last observation. For three of the four groups,

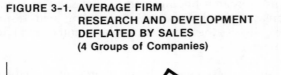

FIGURE 3-1. AVERAGE FIRM
RESEARCH AND DEVELOPMENT
DEFLATED BY SALES
(4 Groups of Companies)

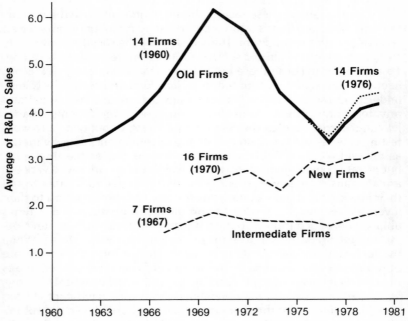

Source: L.J. Butler and B.W. Marion, *An Economic Evaluation of the Plant Variety Protection Act* (Madison: Department of Agricultural Economics, University of Wisconsin, 1983),
Table 3-1.

the trend line tracks the observations fairly closely, but for the oldest
group there is a big bulge in the 1965–75 period. Research reached
a peak relative to sales in 1970, the year PVP legislation was passed,
and declined thereafter. Some thus argue that PVP did not result in
more research.

This argument has major flaws in its use of sales as the deflator.
First, gross sales include the crop value of the seed. When agricultural
prices rise rapidly and are high, as was the case through the 1970s,
gross sales will also rise because the cost of new materials to the seed
company increases. The seed company pays the contracting farmer
who grows seed a multiple of the commodity price of the crops, and
when this rises, it pushes up seed company costs without necessarily
raising net income or profits after taxes. Therefore, part of the decline

after 1970 is due to escalating costs of producing seed and does not reflect either higher profits or larger volumes of sales. The second flaw in using current sales to deflate current R&D is the lag structure inherent in plant breeding. One would expect R&D costs to rise relative to sales as research programs are expanded, but once research begins to pay off, sales would be expected to grow faster. Thus, the ratio of R&D to sales will rise, then fall with no diminishment of R&D effort! The third flaw is much more fundamental—it is the confusing of inputs and outputs. The input into the process is the research expenditure, and the output is the improved performance of the new varieties created, i.e., the value added by the research program. Thus, research expenditure is the appropriate measure for input into R&D efforts, but sales are the wrong deflator because sales measure the output of the R&D program. Where R&D is successful and generates a superior variety, sales would be expected to rise because of higher unit prices and the fact that farmers switched their purchases from other companies. But with output as the denominator, success in research reduces the "measured" level of the research program. This indicator thus divides inputs by a deflator that includes the output of research and is wholly inappropriate.

The correct deflator is an index of the costs of plant breeding research programs, of which the key component would be salaries for breeders and technicians. In the absence of that, the consumer price index is probably as good as any readily available alternative. Results of research expenditures deflated by the CPI are shown in Figure 3-2 for the same four groups of firms. The result is very rapid growth among the largest firms from 1965 to 1970 and moderate growth thereafter, with a departure from the trend in 1977. In this estimate of real R&D growth over time, there is no downturn, but rather a moderation from very high rates of growth in the late 1960s. A further breakdown of research expenditures reported by Butler and Marion indicates that the surge in the late 1960s was heavily weighted to constructing and equipping research facilities while the expansion of staff came in the 1970s. The established companies may have been anticipating PVP, and the attraction of new firms to research in plant genetics throughout the period is probably related to PVP, but neither are established facts. However, the data clearly refute the contention that private R&D in the United States declined after the introduction of PVP.

Although the large firms dominate the total picture, the data in Figure 3-1 show that the newest entrants tend to be as research-intensive relative to firm sales as the older firms. Whether this trend is attributable to PVP is not yet known, but it certainly is the objective of PVP to bring new research-oriented firms into the picture.

The real evidence as to the effect of PVP on research into plant breeding is obscured by the aggregate data; a breakdown by crop is

FIGURE 3-2. R&D EXPENDITURES PER FIRM
All figures adjusted by the CPI (1967 = 100)

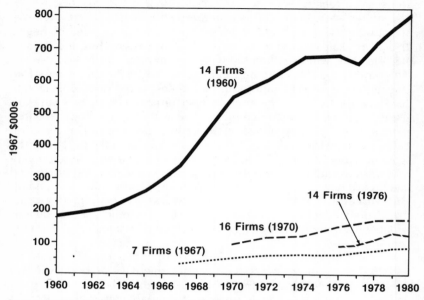

Source: Butler and Marion, *An Economic Evaluation.*

needed. Two recent studies surveyed different samples of firms that break down research expenditure by crop.[39] The results of both, shown in Table 3–8, confirm a basic pattern that is obscured by the aggregate data. Corn is, in fact, the dominant factor in research as well as in sales and value added. Corn declined as a proportion of total research through most of the 1970s, but may have bottomed out or trended upward at the end of the decade. Alfalfa followed a similar pattern, beginning at a lower level, while grain sorghum declined steadily throughout the entire period. This is not to say that research diminished in these crops, but that it did not grow as fast as the surveys' estimates of total research. Both surveys confirm that private sector research in soybeans grew explosively from very low levels to become, by 1980, the second highest single crop in terms of research expenditure. The two surveys also indicate that the greatest part of this growth took place in the 1970s. This would seem to confirm the observation made earlier that soybeans were viewed as the crop with the greatest amount of potential for new breeding effort, once PVP gave private breeders

TABLE 3-8. R&D BY CROP

R&D Expenditure	Current	Percentage of Total R&D Expenditures						
		Corn	Sorghum	Soybeans	Cereals	Grasses	Vegetables	Other
Perrin et al. (59 Firms)								
1960	$ 3,572	52.4	12.5	0.0	0.2	7.2	27.4	0.2
1965	5,707	47.5	11.6	0.6	5.1	10.0	24.6	0.4
1970	11,293	43.5	10.6	2.4	9.6	9.5	22.3	2.0
1975	24,148	42.3	7.2	8.6	12.9	7.5	17.5	4.1
1979	42,630	46.3	6.6	10.1	10.2	7.2	17.6	2.1
Butler and Marion (14 Older Firms)		Corn	Alfalfa	Soybeans	Wheat	Cotton		
1970	3,417	79	9	1	7	4		
1972	3,943	77	8	2	9	5		
1974	5,593	25	7	6	8	4		
1976	8,390	23	6	7	11	3		
1978	10,261	71	6	10	9	4		
1980	12,131	71	7	11	9	3		

Sources: Richard K. Perrin, K.A. Kunnings and L.A. Ihnen, *Some Effects of the U.S. Plant Variety Protection Act of 1970* (Raleigh: Department of Economics and Business, North Carolina State University, 1983); and Butler and Marion, *An Economic Evaluation*.

legal protection for the varieties they developed. Wheat (cereals) had a surge in growth between 1965 and 1975, but has since leveled off.

Soybeans emerge as the single crop that unquestionably has experienced expanded research, as a result of both PVP and the potential seed market. The critical question then is not whether PVP can increase research in one or two attractive crops, but whether research expenditures can be converted into better varieties, increasing market share and revenues to fund future research.

With the expected time lag of 8 to 12 years, new soybean varieties are emerging. However, the private sector's share of this market is still low, and unless it rises, the research cannot be deemed a true success. In this regard, it is worth quoting the president of one of the firms that has made a large investment in developing new soybean varieties.

> Our practical experience is that a farmer will not pay $1.00 more for $1.00 in value. In fact, our experience in seed and animal health products suggests farmers will not buy much of anything unless the return is $4.00 for $1.00 invested; and, if you want rapid impact and big market penetration, you should have a 6 for 1 R.O.I. Our soybean variety A3127 is 6.5 bu./acre better than Williams [the most widely used public variety of soybeans] in the Midwest area of adaptation. Our net price of $14.00—$6–$8 more than grain value— provides a farmer approximately a $6.00 return for $1.00 invested. We have increased sales each year for the four years of sales but still have less than 1/10th of the Williams' acreage.[40]

There are benefits to the consumer in terms of lower prices for food. The seed company is capturing on the order of 15 to 20 percent of the estimated social rate of return to improvements in seed. In such a situation, the tendency to underinvest is persistent and while plant variety protection can help, it is not a definitive solution.

If, in the next 5 to 10 years, the seed industry demonstrates success in converting soybean research expenditures into better varieties, increasing market share and generating income to conduct further research, PVP will have succeeded in its ultimate aim: to bring increased private sector resources into plant breeding and to maintain yield gains even as public funding declines in real terms.

FOOTNOTES, CHAPTER 3

1. Edwin M. Kania, Jr., *Pioneer Overseas Corporation Case Study* (Cambridge: Harvard Business School, Case 4-583-070, 1982).
2. Figures for CGIAR funding received in personal communication with CGIAR official.
3. Pat Ray Mooney, "The Law of the Seed," in *Development Dialogue*, Vol. 1-2 (Uppsala, Sweden: Dag Hammarskjold Foundation, 1983), p. 84f.
4. Ibid., p. 85.
5. See N.W. Simmode, ed., *Evolution of Crop Plants* (Longman, 1976). See also Stephen Smith in *Diversity* (1), 1984.
6. Personal communication with Dr. William L. Brown, former Chairman of the Board, Pioneer Hi-Bred, and world renowned breeder.
7. Bill Paul, "Third World Battles for Fruit of Its Seed Stocks," *Wall Street Journal*, June 15, 1984, p. 34.
8. Personal communication with Dr. George White, USDA, Beltsville, Maryland.
9. Mooney, "The Law of the Seed."
10. Vernon Ruttan, "Changing Role of Public and Private Research in Agricultural Research," *Science* (216), 1982, p. 25.
11. See Vernon Ruttan and W. Burt Sundquist, "Agricultural Research as an Investment," *1982 Plant Breeding Research Forum*, Pioneer Hi-Bred (Des Moines, Iowa, 1982), p. 61.
12. Ruttan, *Science*, p. 23.
13. William Pardee, "Blue Ribbon Panel on Research," *Proceedings of the 28th Annual Farm Seed Conference*, 1981, p. 9.
14. Ibid.
15. Ruttan, *Science*, p. 24.
16. David Padwa, "Public and Private Agricultural Research," *Proceedings of the 28th Annual Farm Seed Conference*, American Seed Trade Association (Washington, D.C., 1981), p. 28.
17. Ruttan, *Science*, p. 24.
18. Ibid., p. 26.
19. "Findings," *1982 Plant Breeding Research Forum*, Pioneer Hi-Bred (Des Moines, Iowa, 1982), p. 35.
20. Robert F. Leibenluft, *Competition in Farm Inputs: An Examination of Four Industries* (Washington, D.C.: Federal Trade Commission, 1981), p. 86.
21. Mooney, "The Law of the Seed."
22. ASTA analysis of sources cited in *Seeds of the Earth*.
23. L.J. Butler and B.W. Marion, *An Economic Evaluation of the Plant Variety Protection Act* (Madison: Department of Agricultural Economics, University of Wisconsin, 1983), Appendix A, p. A-6.
24. Ibid., p. 42.
25. Leibenluft, *Competition*, p. 98.
26. Kania, *Pioneer Overseas Corporation*, p. 9.
27. Ibid., p. 30.
28. USDA Crop Reports.
29. FAO, *Annual Yearbook on Agriculture*, 1982.
30. See results of a survey by Butler and Marion, *An Economic Evaluation*, p. 71.
31. Ibid., pp. A-2 and A-3.
32. Mooney, "The Law of the Seed," p. 152.
33. Butler and Marion, *An Economic Evaluation*, p. 72.

34. Louis W. Goodman, with Arthur L. Domike and Charles Sands, *The Improved Seed Industry: Issues and Options for Mexico* (Washington, D.C.: Center for International Technical Cooperation, 1982).
35. Butler and Marion, *An Economic Evaluation*, p. 72.
36. Ibid., pp. 3–4.
37. See, for example, Mooney, "The Law of the Seed," pp. 153–158.
38. Ibid., pp. 153 and 154.
39. Richard K. Perrin, K.A. Kunnings and L.A. Ihnen, *Some Effects of the U.S. Plant Variety Protection Act of 1970* (Raleigh: Department of Economics and Business, North Carolina State University, 1983); and Butler and Marion, *An Economic Evaluation.*
40. Letter from John Studebaker, President, Asgrow Seed Company, to Richard Perrin, November 2, 1982, as quoted in Butler and Marion, *An Economic Evaluation*, pp. 85–86.

The Internationalization of the Seed Industry

4

INTRODUCTION

This chapter deals with the spread of late vintage improved seed to new markets in the developed countries. The principal focus will be on the United States and Western Europe because they are the leading producers and exporters, and also because data on their trade flows are available. However, the subject is broader than international trade flows; the internationalization of the seed industry includes local production, overseas production and trade of improved seed. The theory of international trade assumes the mobility of goods internationally and the immobility of factors of production (i.e., the labor, capital, land, and technology that combine to produce the goods); hence, the usual emphasis on exports and imports. Available data indicate that exports of seed have been growing much faster than agricultural output for at least two decades. However, statistics on seed entering world trade understates the true extent of the internationalization that is taking place. Farmland and farmers are factors of seed production that are not mobile internationally, but plant breeders, seed production specialists and technology are mobile and have helped to bring superior seed to new markets. Because of natural and legal barriers to trade and various advantages to producing locally, many companies prefer to import into the overseas market only what they need as parent seed and to produce commercial seed locally. Other firms have no choice in that local conditions require specific types of seed that can only be developed and produced locally. The result is that farmers around the world are benefiting from internationally improved seed through the transfer of knowledge, even though the improved commercial seed may not appear in the trade statistics.

Just as the seed industry is in a period of transition, so too is international trade in seed. Historically, seed has been a commodity in which trade was not large relative to overall production. The seed industry was fragmented and dominated by small local firms until the 1970s. There was also the prevailing view that variation in local conditions, specifically temperature, moisture, day length, soil chemistry, pests, and diseases precluded significant trade in seed. Because of the structure of the industry and problems in developing widely adaptable varieties, the seed industry did not perceive the true possibilities for growth in international trade until the 1970s. The worldwide impact of high yielding varieties of rice and wheat helped to change this perception, and it is now clear that the potential for the expansion of international trade in seed and the mutual benefits flowing from it are enormous.

Overseas production by seed companies was also rather limited before the 1970s. Firms utilized foreign growers for climatic reasons. German seed companies, for example, have a history of multiplying seed in other countries for sale in Germany. The reason is that the relatively cold climate in Germany is not conducive to seed multiplication for some species of plants. Therefore, production of the seed was contracted out to farmers and growers in more suitable locations, often in Eastern Europe, and commercial seed was then imported into the German market. There was, however, very little large-scale research and production effort by seed firms to gain access to distant markets. Neighboring markets were developed with sales among European nations sharing somewhat similar conditions, and the United States sold to Canada and Mexico. But such activities were more the byproduct of domestic sales than a conscious effort to expand overseas. An important obstacle to establishing seed production operations overseas was the variation in legal treatment and protection among countries. Before plant variety protection was established and standardized internationally in the form of UPOV, firms were hesitant to set up the research, production and distributional operations needed to achieve success in a new market. Without reasonable legal protection, the necessary investment could not be justified. Even in the developed countries, problems certainly still exist with the legal and regulatory regimes governing seed, but the situation is far better than it was in the 1960s. As a consequence, in the 1970s seed firms began to emphasize the expansion of research, production and distribution activities in overseas markets.

Before 1970, public research agencies were often instrumental in the introduction of improved foreign varieties. This is true in the United States and in France and also among developing countries where high yielding wheat and rice have been successful. This was generally based on crosses between high yielding foreign varieties and suitable locally adapted varieties. In recent years, however, private seed companies have become more active in spreading new varieties of improved seed around the world, principally hybrid corn and sorghum. While public agencies continue to play an important role, the initiative in introducing new varieties of improved seed for the major crops is now with the private sector. This is part of the restructuring of the seed industry described in the previous chapter and has important consequences for seed production and use worldwide. When new seed was introduced by public agencies, either the international centers in the developing countries or national research centers like INRA in France, they typically worked through existing organizations. This had the effect of sustaining the established structure of the seed industry or, at the most, of reordering the position of local organizations and firms. Publicly introduced seed was not a vehicle for significant structural change in the local seed industry and, in particular, did not encourage local firms to make serious efforts to upgrade their research capabilities. This began to change in the 1970s, and by the early 1980s, it was possible to identify some of the likely consequences of the introduction of internationally improved seed into new markets by private firms. Some of these results followed from earlier public introductions of improved seed, but others were new developments. These consequences include:

- rising agricultural yields with positive effects on farmers' incomes, consumers' diets and the nations' balance of trade[1];
- diminished market shares for local seed producers, often resulting in pressures to reduce competition through barriers to trade and regulations on production;
- increased dependence on imports and foreign suppliers of seed;
- need for expanded and improved research programs by local firms if they are to remain competitive with the new varieties introduced by overseas firms;
- greater need to conserve existing genetic resources as farmers switch from diverse low yielding local varieties to the more uniform, higher yielding international varieties.

This list does not exhaust the effects of the internationalization of the seed industry, but it indicates some of the adjustments that are under way and likely to continue for at least the next decade. The internationalization of the seed industry, through trade and overseas production, reinforces the trends toward concentration and privatization discussed in the previous chapter. Both domestic and international forces are working together to restructure and internationalize the seed industry. This process is at different levels in different countries. In some countries, notably in Western Europe, these forces are well advanced, while in other regions of the world they are just beginning to impact on local seed industries. The next section of this chapter will look at the motivations and objectives of importers and exporters. The final section will review trends and patterns in the trade data for Western Europe and the United States.

INTERNATIONAL TRADE IN SEED

Traditional analysis of international trade utilizes the theory of comparative advantage as its conceptual basis. Countries will tend to export those products in which they have the greatest comparative advantage and import goods for which they have a comparative disadvantage. A country's comparative advantage would manifest itself in the production cost and/or quality of the products produced. Where a country is a high cost (and/or low quality) producer, it does not have a comparative advantage and should import. The determinants of comparative advantage are complex, but can be explained in terms of factor endowments. Different goods require varying amounts and types of factors of production, and countries are endowed with different factors of production. Australia and Argentina have land, Japan and Germany have capital and skilled labor, Saudi Arabia has energy resources, and so on. Production processes are somewhat flexible, but not sufficiently so to make Germany or Japan exporters of agricultural goods or raw materials. There are several problems associated with this approach, but as a start in explaining trade flows, it is a plausible basis.[2]

The United States is a difficult country to categorize in terms of factor endowment because it has a great deal of good land, a skilled labor force and plentiful capital. Over the years, the comparative advantage of the United States has tended to shift as world trading patterns have evolved. The United States emerged from World War II with a strong competitive position across the board. As the European and Japanese economies recovered in the 1950s and 1960s, American comparative advantage emerged in knowledge and skill-intensive products. Through the 1970s, imports made inroads into such capital-intensive industries as steel and automobiles. In the 1980s, most analysts would tend to agree that the U.S. comparative advantage lies in agriculture, high technology products and knowledge-intensive services. The seed industry is particularly interesting because it combines aspects of all three areas and, not surprisingly, the United States is the leading net exporter of seed in the world.

Seed Exports

Looking more carefully into the concept of comparative advantage, it is possible to identify several specific characteristics that enable a country to export seed. In large part, these characteristics are the availability of factors of production that are essential to produce seed, as theory would suggest. But certain institutional factors are also important, and there is a fundamental prior condition — a domestic market for high quality seed that encourages the prior development of a strong local seed industry. In considering the countries that have had sustained success in exporting seed, the following conditions seem to be important:

(1) a relatively strong domestic agricultural sector that can af-

ford and productively utilize high quality seed;
(2) climate and latitude (day length) conditions suitable for multiplying seed;
(3) seed production capabilities ensuring secure, high quality
supplies of seed at competitive prices;
(4) a sizable domestic base in agricultural research, including
active public and private plant breeding programs;
(5) a sufficient pool of plant scientists and an educational system that prepares a sufficient number of well trained new breeders for public and private genetic research;
(6) domestic laws protecting property in general and plant varieties in particular that enable successful investors and breeders
to benefit from the contributions they make;
(7) an active and effective seed certification organization to
ensure that basic quality standards are met by all seed producers;
(8) the evolution of a domestic seed industry with a number
of large integrated firms and agencies capable of research, production and distribution;
(9) the cumulative buildup of genetic knowledge that is applicable directly (exports) or indirectly via plant crosses (overseas
operations) to the seed requirements of foreign countries.

This listing begins with the initial, most basic conditions and enu-
merates subsequent levels of seed industry development that generally
have to be met if a country is to be a sustained exporter of seed. A
country that fulfills all these conditions is almost certainly going to
be a successful exporter of its own seed. However, it is possible for
a country to have a strong research base and breeding expertise, but
to be uncompetitive in production. In this case, the actual production of seed is likely to shift overseas to the country that will ultimately
use the seed, or to a suitable low cost seed producing country in the
region, while the research effort is still centered and managed domestically. Logically then, it is possible for a particularly good multiplier
of seed (having good climate and efficient production) to export without having a strong research base. As specific cases are discussed in
this chapter, examples of these types of countries will become evident.

Some of the nine conditions fall into the category of traditional
factor endowments: climate, plant scientists and farmers capable of
efficient multiplication. Others derive from institutional arrangements:
breeder education, legal protection, effective certification programs,
and the growth of integrated seed firms. These prior conditions interact
to create a situation in which exports can take place; that is, the domestic seed industry has built up knowledge that can be applied to
improve crop yields overseas. The product ultimately being exported

is the genetic knowledge embodied in improved seed; the seed is only the conveyor of accumulated genetic knowledge. The theory of comparative advantage is not dynamic in that it does not explain why or how export patterns are likely to shift over time. Neither does it explain how factor endowments are likely to change nor why a country's export mix can shift without a change in factor endowment.

One of the best dynamic explanations of trade patterns is the product life cycle theory.[3] This explanation begins by observing that products go through a cycle of being introduced as new, often high technology goods, become more common after a number of years, and ultimately are produced as fairly standardized low technology goods. This also has implications for trade when the product is initially exported by the country developing the new product and ultimately ends up being imported. Instances of product life cycles abound, with the production of radio receivers and other simple electronic components furnishing recurring examples. When vacuum tubes were on the cutting edge of technology, the United States and a few other advanced countries produced and exported radios. In time, these production processes became well known, and Japan and other countries took over the production of vacuum tubes and related electronic equipment. Subsequently, the fabrication of these products became highly standardized and was performed in various developing countries. About 20 years later, the process was repeated with transistors as the basic element of the electronic component. More recently, printed circuits and silicon chips have emerged as high technology in electronics, and the process may be set to run its course again. Many products once thought to be sophisticated manufactures — and the preserve of the advanced industrial countries — are now produced competitively by newly industrializing nations in the developing world and exported to Europe and the United States. Television sets are an example. Steel has moved along the cycle, and automobiles may also follow this pattern. Competitive final assembly of autos shifted some years ago to include Japan; now Korea is trying to break into the international market. Subassemblies and components are widely manufactured in countries such as Brazil, Mexico and Taiwan.

The product life cycle theory can be applied in varying degrees to the seed industry. Particular varieties certainly follow the cycle, beginning as a new high yielding breakthrough representing high value added by the firm that developed it. In time, particularly for open-pollinated seed, others begin to produce the new variety — perhaps under a different name or for personal use only — and the variety becomes a mature product. Eventually the variety is known all over the world and produced locally wherever there is market for the seed and plant variety protection, or the realities of the situation allow. Another process is at work among hybrids to make today's leading variety tomorrow's standardized product; that process is competition through research and the introduction of better varieties. It is very unusual for a hybrid to remain the leader in a particular market for more than six or seven years. And many new varieties become mature or obsolete in shorter periods as competing plant breeders introduce superior products. Therefore, particular varieties of seed are subject to product life cycles, and they can move from the status of new technology to maturity to standardization or even obsolescence fairly quickly. Furthermore, as a particular open-pollinated variety becomes a better known standard product, it will likely be widely produced by farmers for personal use and by local seed firms for sale in countries that lack plant variety protection. For an archetype following this "plant variety" life cycle, the first stage would be introduction as a new improved high yielding variety for the domestic market. The second stage would be increased exports to suitable regions as domestic demand becomes satisfied and supply capacity continues to expand. The third stage would be overseas production using lower cost sources near the ultimate market for seed multiplication. The fourth stage would occur when the originator phases out the variety and small-scale local producers meet whatever demand remains. The time sequence of this cycle can vary widely, depending upon how fast breeders develop new and better varieties to supplant the existing market leader. This sequence implies that the variety has wide geographic application. Where this is not the case, the same process could follow if the new high yielding variety is bred with an appropriate local variety to produce a high yielding variety adapted for local conditions.

While it is useful to describe the full life cycle of a commercially successful plant variety, this pattern is certainly not universally followed. Successful domestic varieties often are not suitable for direct use abroad, and the export stage may be skipped entirely. More and more frequently, an alternative stage occurs where a special breeding program is set up to adapt successful domestic varieties to conditions abroad. An overseas breeding program may be utilized to cross the domestic variety with local varieties to develop seeds that are adapted to foreign agricultural conditions. Once suitable varieties have been developed, they are usually multiplied in the foreign country for ultimate sale as commercial seed to local farmers. This approach follows the pattern that worked for the international centers in spreading high yielding wheat and rice varieties to many developing countries. These successes showed that the benefits of genetic research could be shared more widely throughout the world, and the more dynamic private seed companies have set up research centers in various regions to develop and to utilize their genetic knowledge as broadly as possible.

While particular plant varieties inevitably go through product life cycles, it is not established that the leading seed—or seed exporting—companies follow this pattern, as occurs in many other industries. There are barely observable trends indicating that this might take place, but no conclusions can be drawn. The situation is made more complex by the absence of adequate plant variety protection in many countries and the obstacle this creates in developing a strong local seed industry with potential to export. As noted above, the most important item being produced and exported by the seed industry is genetic knowledge embodied in seed, and as long as companies continue to produce knowledge of value to agricultural producers overseas, they should be able to continue to export. The product cycle for particular varieties can be very short, so research, breeding and the introduction of a continuing succession of improved varieties is the only way to maintain the loyalty of farmers and a reputation for high quality of seed. When a seed company becomes too dependent on a single variety and does not renew its product line, it runs the risk of being overtaken by others with better varieties. This is the inevitable fate of any seed company that views its products as particular lines of seed, rather than as the embodiment of genetic knowledge. Firms that hold this static perception of the seed industry are likely to follow closely the product life cycle of their current varieties. A dynamic perception of the seed industry recognizes that knowledge is the real product and research the key to meeting the needs of farmers and retaining a position of leadership.

This line of reasoning supports the view that, while particular varieties are subject to obsolescence and decline in patterns somewhat similar to the product life cycle observed in other industries, there is no reason why seed companies should follow such a pattern. If companies are successful in developing new improved varieties to replace existing products, they should be able to maintain their standing as suppliers of improved seed. The production of genetic knowledge is by definition high technology and should not be susceptible to product life cycle patterns. There are few indications to date that genetic research or even traditional plant breeding are becoming standardized operations that can be done more efficiently by lower cost labor abroad. The opposite may be occurring; with new bioengineering breakthroughs beginning to appear, genetic research is closer to the cutting edge of high technology than ever before.

Most leading seed companies have some plant breeding operations in the tropics and Southern Hemisphere to take advantage of climatic differences and to accelerate the development of new varieties. However, these operations are small relative to breeding programs conducted in the United States and Western Europe. These programs are primarily used to support and accelerate breeding efforts in the home country and their results are integrated into companywide research efforts. These overseas operations indicate the current potential for companies conducting research abroad to take advantage of differences in climates. However, it is unlikely that an emigration of genetic research programs for open-pollinated crops is going to accelerate in the next few years. As just noted, one important obstacle is the uncertain status of plant variety protection in many countries. This limits the willingness of companies to invest in new facilities and establish breeding programs, for nonhybrid crops, to serve markets with hesitant and changeable plant protection laws. With improved PVP, there would certainly be more private research done in developing countries, but it would probably not replace research now being done in the United States and Western Europe. It would aim to serve local markets and, at least initially, would borrow knowledge developed by the previously existing research programs. In time, the local research efforts would become more independent as they accumulated genetic knowledge, but they would still be able to draw on relevant breakthroughs made in other countries. Exports to similar markets might also develop as domestic needs are met. These trends are now only barely visible, and it is largely conjectural to project what would happen if seed companies were to make a significant commitment to supply developing country markets. The absence of adequate PVP militates against such a commitment. Therefore, until the legal issues are better clarified, if not resolved, seed companies producing open pollinates are likely to seek a presence in developing countries, but are unlikely to establish significant integrated operations as they have done in Western Europe and the United States. For hybrids, the establishment of local research programs aimed at sensing local needs would advance more rapidly. However, as will be demonstrated in the next chapter, public seed policy is all important.

While the production of genetic knowledge is unlikely to depart significantly from the advanced countries in the foreseeable future, there is some indication that the production of seed embodying that knowledge is more mobile. Although aggregate export figures for the United States and Western Europe still show strong growth, some seed multiplication activities will move to countries and regions that can maintain quality standards, but have lower production costs and perhaps climatic advantages. An example of this is the increase in corn seed imported into Western Europe from Eastern Europe in recent years. Lower production costs in Hungary and Romania are a factor in this trend, but the other side of this issue is the high seed production costs in the European Economic Community resulting from the Common Agricultural Policy. This will be discussed in more detail, but suffice it to say that the CAP creates a market distortion that artificially raises seed production costs in Western Europe. As a result, France—the major seed producer in Europe—is becoming less price competitive in the production of corn and other seeds that face this problem. Therefore, it is possible for countries inadvertently to push seed production abroad through market-distorting policies that raise seed costs. In this way, a country can create or accelerate a product life cycle pattern for the multiplication of seed even where domestic research efforts are strong. The evidence, therefore, indicates that, while the leading countries and companies do not face a near-term competitive threat in terms of producing knowledge, the actual multiplication of seed can move to new areas if comparative production costs are out of line. Seed producers in the United States were facing a cost threat in terms of the unrealistically high U.S. exchange rate prevailing in the mid-1980s. This raised American production costs relative to foreign producers and almost surely reduced exports in this period relative to what they could have been.

Concerning the most important product of the seed industry— genetic knowledge based on plant breeding research—the advanced countries, particularly the United States, appear to have a strong and growing comparative advantage. In basic and applied research, in plant biotechnology and in traditional plant breeding, this position seems assured through the 1980s and probably beyond. In traditional plant breeding, the developing countries are building up their capabilities, but they are fully occupied with local needs, and exports are merely a felicitous byproduct when they occur. The great question for the developing countries is how to best tap into the biotechnology and genetic knowledge that is growing throughout the world. This will be discussed in Chapter 6.

With regard to the production of seeds themselves, this activity could begin to migrate from the current exporters—Western Europe and the United States—before the end of the decade. Seed multiplication costs are rising in these areas, and unless the European market distortions and problems with excessive dollar volatility are corrected, some erosion in their export position is likely to become evident. Seed production will probably not run the full product cycle and be exported back to the United States, but in Western Europe, favorable climate and cost advantages in Eastern Europe make this a distinct possibility. This is most likely for high value added crops such as corn and sorghum. For other crops, such as wheat, the low value added relative to shipping volume and cost makes the shift of seed production to overseas sites more problematic.

Seed Imports

In terms of commercial seed inputs, the theory of comparative advantage holds that when a country is not adequately endowed with one or more of the necessary factors of production, it will be a high cost producer and should import at least part of its seed requirements. The reasons for importing can be described in more detail than this, however. Just as with exporters, importers of seed tend to be characterized by several conditions; moreover, their import policies indicate that importers are motivated by a set of goals that they try to achieve through management and, in some cases, control of imports. Importers of commercial seed tend to be characterized by the following conditions:

(1) at least part of the domestic agricultural sector is relatively advanced and able productively to utilize improved seed;

(2) quotas and other legal restrictions on imports are not so pervasive as to restrict trade entirely;

(3) climatic conditions may not be suitable for multiplying all of the various seeds needed by domestic agriculture;

(4) domestic production of high quality seed at competitive prices is inadequate;

(5) the domestic research and technical manpower base is not sufficient to meet the full range of the country's plant breeding requirements; or

(6) the legal environment may not encourage the development of a private seed industry.

For imports to occur, the first and second conditions must apply, along with one or more of the remaining conditions. Initially, there must be a need for improved seed in terms of local farmers who can combine them with fertilizer, water and other necessary inputs, producing a sufficient increase in yield to make a profit over and above the cost of the new seed and other inputs. Agriculture at its most basic subsistence level is not in a strong position to utilize improved seed profitably. Rural and agricultural development, including infrastructure (farm to market roads, for example), training and provision of complementary inputs, must precede or at least accompany the introduction of improved seed. Without supporting development, the introduction of improved seed is unlikely to generate consistently the yields necessary to justify the increased costs. Therefore, demand for imported seed requires at least some initial success in domestic agricultural development. Likewise, commercial seed imports cannot flow if they are legally barred from entry, as is the case in many countries. This subject will be examined in Chapter 8. Suffice it to say here that many developing countries severely restrict seed imports and, as a consequence, end up losing agricultural output and foreign exchange within a two- to three-year time frame. Not only must domestic demand exist, but it must be free to enter international markets to satisfy its seed requirements in excess of domestic production. These are necessary conditions for imports; beyond them, it is sufficient that one or more of the other conditions pertain. However, where a country is a significant net importer of seed, it is usually due to one or more of the other conditions being present. Except for climate, all of the conditions noted are conducive to remedy through corrective public policy, and it is unusual for a country to allow just one of these conditions to stand in the way of greater self sufficiency in seeds. It is commonly the case among significant importers that several of these conditions prevail simultaneously.

It should be noted that all countries import some amount of commercial seed and, for most countries, imports of certain types of seed proceed on a regular basis. Even the leading seed exporters, the United States and France, are also significant importers of seed. As noted, the United States has developed a strong comparative advantage in seed production and exports a broad range of crops and varieties, but is still a sizable commercial importer of several types of seed, including alfalfa, legumes and numerous garden and flower seed. Some of the flower and garden seed is contracted out by U.S. firms for multiplication overseas. A small country like the Netherlands has developed a comparative advantage in a few crops and is a big exporter of these, while importing other types of seed. At the other end of the spectrum, countries such as Japan, Germany and Switzerland export very little in the way of seed and meet a large proportion of their domestic requirements through imports. For most of the developed countries, seed imports are relatively free, and the trade patterns reveal the underlying structure of domestic supply and demand. For most developing countries, this is not true, and imports are repressed through various barriers to trade. As a result, it is more difficult to speak with assurance regarding the actual balance of internal supply and demand in developing countries.

In recent years, the debt problems of several of the leading developing countries in Latin America have restricted foreign exchange availability, and purchases of seed, together with many other imports, have been reduced. Seed exports from developing countries are generally small, where they exist at all. Some of the developing countries have enormous potential for seed production and, ultimately, exports. But they must first overcome certain problems. The specific challenges are in building up a research and technical manpower base, developing an efficient, integrated seed production industry, creating a legal climate that will foster the growth of a private domestic industry, and working out a mutually beneficial relationship with the leading international seed companies.[4] The availability of efficient low cost labor and the appropriate climate has led firms from the United States and Europe to contract out for seed multiplication in certain developing countries; examples are flowers in Kenya and tomatoes and garden vegetables in Taiwan. The propagated seed is returned to the developed country for sale in the domestic market and, in some cases, for export. Such arrangements are the first stage in the development of a comparative advantage in seed production by developing countries and benefit all involved.

The analysis up to this point should not obscure the reason that firms and individuals enter into foreign trade; they aim to cover their costs and earn an income. At the level where most trade takes place, the primary motive of earning a living and making a profit has to be recognized. Without mutual benefit for both parties, there will be no transaction. International trade in seed is not a zero sum game in which gains on one side automatically imply losses on the other. As noted in the previous chapter, even after all costs are taken into account, farmers in the United States tend to benefit roughly fourfold from their purchase of new seed. Similarly, when accompanied by an appropriate program of agricultural development, the introduction of improved seed in the Third World has had, and will continue to show, gains that far exceed costs.

While increased income is the motivation that influences those directly involved in international seed trade, the motivations influencing government policymakers are more diverse. Many governments discourage seed imports on mercantilist or security grounds. They do not want to spend the foreign exchange or become dependent on foreign suppliers for something as basic and essential as agricultural seed. If the country's agricultural system is sufficiently advanced so that it could benefit from imported improved seed, then the mercantilist argument against imports is totally incorrect. In many cases, the cost-benefit ratio is sufficiently high in a time interval of a year or less that the gain in foreign exchange from higher yields exceeds the cost of the seed. The security argument is more complex and could cut both ways. Dependence on foreign suppliers per se does not increase a country's insecurity; most would argue that a diversity of suppliers will reduce insecurity. Problems can arise when the imported seed is not adequately tested and has an unanticipated weakness or when the indigenous agricultural system is unprepared to provide complementary inputs at the right time and in the right amounts. These problems can be reduced by a thorough testing program for genetic suitability. This should be followed by the slow introduction of new seed to a sample of the farmers most likely to be able to utilize its potential. Then, depending on these initial results, the rate of introduction can be accelerated or slowed. In this way, risks are minimized, and the agricultural system can evolve to supply the new inputs required in conjunction with the imported seed. These are problems that are manageable, and the security argument is not definitive.

In analyzing seed policies in countries that import seed on a regular and sustained basis, several underlying goals emerge, and it would be worthwhile to consider them explicitly. Taken together, the goals represent the objectives of a good seed import program and can be grouped under five general headings: balance of payments, security of supply, price competition, seed quality, and access to improved germplasm. It is the rare case when an importing country can achieve all the objectives simultaneously, but the realization of three or four of them is typical.

Before proceeding to examine these in more detail, it must be reemphasized that there are critical physical limitations to seed imports that result from the inherent botanical properties of the plants. For a whole series of reasons, seed must be produced semi-locally or at least in comparable climates in other regions. Day length sensitivity is probably the major factor limiting the location of seed production. Grasses are particularly sensitive to day length as are most varieties of corn, sorghum, rice, and wheat (which, strictly speaking, are species of grass). Therefore, seed must be produced at the same latitude as the region in which it will be eventually used. Vegetables are generally not as sensitive to day length; thus, there is more scope for trade flows. Other important factors limiting the regions in which seed can be multiplied are the availability of moisture—both ground moisture and humidity for some types of crops—and diseases and pests. All these factors combine to limit naturally the scope for international trade in seed and to make local production the dominant source of supply for most countries. However, where there is scope and need for imports of seed, international trade can make a valuable contribution to agriculture in the importing country.

Seed imports can be viewed as a substitute for food imports, or a means to expand food exports, and therefore as a vehicle for saving foreign exchange, improving the balance of payments, and increasing domestic output in the short run, i.e., one to two years. Assuming that domestic agriculture has advanced to the point where it is able to utilize improved seed, the yield gain from the importation of seed and other required inputs should save foreign exchange, compared with importing food. Moreover, the effect should be felt within one or two crop seasons, meaning the payoff is almost immediate. If some of the other inputs can be produced domestically, the foreign exchange savings would increase. The most important caveat, of course, is whether local farmers can master the new technology inherent in using improved seed and achieve the yield increases that are expected. Another consideration is whether the food imports being replaced by imported seed are concessionary, i.e., being supplied at below market prices. The calculation would have to take into account the actual foreign exchange cost of acquiring the seed and other imported inputs compared with the actual savings resulting from reduced food imports. Where the seed does not substitute for imports, but will be used to improve local diets, the availability of foreign exchange must be considered as well as the alternative uses to which the foreign exchange could be put. Evaluating the importation of improved seed is not difficult and almost inevitably makes sense providing that the farmers can properly utilize the seed, and that the output replaces food imported at market prices or exported and sold overseas at market prices. Under these conditions, there is a strong balance of payments argument for importing improved seed.

Seed imports can be used to enhance the security of supply by maintaining working relations with reliable sources of imported seed to supplement domestic production. Weather is the most obvious variable, but seed production—as with agriculture in general—has greater inherent unpredictability than most other production processes. Occasional shortfalls occur in all regions and countries, and it is important to have alternative suppliers that can be counted on when domestic production or a foreign supplier is unable to provide the customary volume and quality of seed. Seed security can also be enhanced by importing a diversity of varieties with different genetic backgrounds. Such a policy will reduce the damage that a disease or pest can do to the country's agricultural production. If one variety is damaged and others are unaffected, it is relatively easy to solve the production problem by switching from the vulnerable strain. It also helps breeders correct the genetic weakness in the heavily affected variety by identifying varieties that are unaffected. Imports of genetically similar varieties will not achieve this desired diversity. There is no way to know fully the genetic background of the varieties imported, but reasonable precautions can be taken to ensure that they are different from one another.

Italy faced the problem of large-scale imports of hybrid corn seed in the early 1980s without sufficient genetic diversity. Although marketed under different names, the same variety comprised about two-thirds of the market in the 1982–83 season. This is a case of insufficient diversity and makes the crop at least potentially vulnerable to a single pest or disease. Italian policymakers are aware of the problem and are making progress in solving it. Imports should be a means of improving diversity and security, not increasing genetic risks by making a country overly dependent on one or two varieties.

Seed imports can be used to increase price competition and to ensure that farmers have a wide choice of seed at prices at or below those prevailing in world markets. If a country limits its seed suppliers to domestic producers or a few traditional suppliers, it runs the risk of having to buy high cost seed, putting its farmers at a disadvantage vis à vis those of other nations. Wherever competition is limited by law or tradition, those that have the right to the market begin to get complacent and eventually become high cost producers. In the seed industry, this would be evident in terms of insufficient or inappropriate research and a failure to develop new and better varieties. Under normal circumstances, it takes several years for the decline to become evident, but by that time, drastic action may be needed to correct the situation. Competition is the surest means to offer the agricultural sector an ample choice of competitively priced high quality seed. Competition will help to bring domestic firms' performance up to international standards, and it will prevent traditional importers from treating the market as their captive and raising margins. This is not to argue that a country should immediately throw open its seed market to all competitors. Some countries already are open and should continue to enjoy the benefits of competition; others are not now competitive and cannot open their markets without seriously damaging their domestic seed industry. In this situation, the objective should be increased competition over time, as part of a coordinated plan to improve the local seed industry and to gain access to the accumulated knowledge of international seed firms on a mutually advantageous basis. It is not possible to have a vigorous and dynamic seed industry without competition, and it is the very rare country that can develop sufficient competition among only local firms. On the international level, competition is based on continual research and development and the introduction of improved seed varieties. Imports are a means of tapping into this worldwide pool of genetic knowledge and pushing the local seed industry to higher levels of efficiency and performance while holding prices down to at least international levels.

Imports can be used to improve the quality of domestic seed in terms of yield, genetic purity and consistency, and physical standards such as size, high germination rates and freedom from weeds. Assuming that the imported seed has been properly tested and certified and that the country is dealing with respectable and established seed producers and exporters, then imports will tend to be a guarantee of seed quality. Farmers will become accustomed to better quality and will demand higher standards. This in turn will encourage the local seed industry to maintain or, if necessary, to enhance their seed quality to remain competitive with the imports. As a general rule, imports are not qualitatively superior to local seed in terms of resistance to local diseases and pests because they have not gone through the natural selection process in the local environment over the centuries. This is a potential problem that must be guarded against through adequate certification tests, maintaining a diverse genetic base in the varieties planted, and monitoring results in farmers' fields. But in terms of the physical quality of the seed produced and sold to the farmer, imports can be a help in upgrading the standards of the domestic seed industry.

Finally, and most important, imports can be used as a vehicle for improving the genetic base of the seed used in domestic agriculture. The physical quality improvements that result from imports occur on a one-time basis, and seed must be imported on a sustained basis before local standards are brought up to international quality levels. Beyond this factor, imported seed provides access to improved varieties created by breeders in other countries that may significantly benefit local agriculture. In this way, the genetic quality of domestic seed stocks can be permanently improved. For breeding purposes, imported germplasm is of course an invaluable tool for upgrading domestic breeding programs and improving locally adapted seed. This process was described above, and its importance cannot be underestimated. The key to ensuring maximum benefits to breeding programs around the world is to improve the existing system for conserving genetic resources and to keep open genetic exchange programs between national and international agencies.

Beyond their benefits to breeding programs, imports of improved seed can upgrade the genetic base of a country's seed stocks more immediately and more directly than can local breeding programs. For crops that propagate through open pollination, trade can be a low cost substitute for a research program. By entering the world market periodically and importing appropriate new varieties of improved seed, a country can secure yield boosts and other improvements without having to establish and fund a domestic research program. While admittedly an imperfect substitute, periodic imports are a stopgap method of bringing in improved germplasm for crops without an adequate domestic research program. This will not work for hybrids, as discussed, but is certainly one of the attractions of imports of open-pollinated crops. Once the imported seed is distributed, farmers can save part of their crop for sowing the following year, and in this way the genetic properties of the imported improved seed are available to local agriculture on a continuing basis. By periodically buying the newest varieties from the international seed industry, a country can update and renew its genetic base in open-pollinated crops without having either research or production facilities. When developing countries can find suitable varieties, this approach to genetic improvement is very attractive. Many of them lack the resources to establish their own research and production facilities covering the crops of interest to domestic agriculture. For them, imports can be a means to piggyback on research done elsewhere and achieve periodic improvements in yield and other important genetic characteristics at very low cost. International plant variety protection laws do not prohibit farmers from retaining seed. However, countries that have ratified the UPOV convention are committed to prohibiting domestic producers from multiplying and selling imported seed. This may be a factor in the reluctance of developing countries to ratify international PVP agreements, but no systematic analysis of this is available. Even when international plant variety protection does apply, the potential is enormous for improving the genetic base of a country's seed stock through farmers' actions.

Imports provide countries with several avenues to improve their national seed programs. A well designed import program is an essential part of a comprehensive long-term strategy for expanding and strengthening domestic seed production. Imports of seed can provide the following important benefits:

(1) making accessible new germplasm for domestic plant breeding programs;

(2) saving foreign exchange by increasing agricultural output;

(3) improving the security of seed supplies by diversifying sources of supply;

(4) increasing price competition;

(5) raising standards of the physical quality of the seed;

(6) directly upgrading the genetic base of open-pollinated seed.

It is unusual for all these benefits to occur simultaneously from a particular transaction, but it is possible as none of them are mutually exclusive. It is, however, very unlikely that these benefits will result in the absence of an integrated seed program aimed at building up national capacity, developing a mutually beneficial and continuing relationship with at least two or three international seed companies, and using imports to fill the current gaps in research and production. Imports can also be used to serve notice to national and international firms with local operations that they will have to stay competitive in terms of research and production to maintain their position in the local market. Permanent protection against seed imports provides contrary signals to the local seed industry. If the domestic market—such as it is—can be taken as secure, then local seed producers are less likely to make as great an effort to expand their research and improve their performance. Permanent protection also lessens the chance that local producers will become sufficiently competitive to export on a sustained basis. Partial and temporary protection is different. It can help bring in international seed firms and give the domestic industry time to establish itself. But to have the desired effect, protection must be recognized as temporary, with a pre-established rate of decline, and must be part of an overall national seed program. An enduring pattern of protection condemns a country to enduring inferiority in its seed industry.

EUROPE AND THE UNITED STATES

In this section, the available statistics on international trade in seed will be reviewed and analyzed, focusing on the United States and the EC because available trade flow data exist for most recent years. Unfortunately, information on U.S. imports of seed was not collected after 1981 due to budget cutbacks; however, data for earlier years confirm that U.S. imports were small relative to exports and, in value terms, included a significant share of contract multiplication of seed that was broken out of total imports. The most detailed data are available for American exports, and most of this analysis will concern these statistics. Relative to total U.S. seed production, exports vary considerably in importance. They are a small part of seed production for the staple U.S. food crops; exports comprise only 4.0 percent of corn seed and 4.3 percent of soybean seed. But for other U.S. seed producers, exports are a significant part of the overall market, totaling 50 percent of vegetable seed and 36 percent of grain sorghum seed.[5]

The United States is a net exporter in all seed categories. Thus, in comparison with all traded goods (industrial and agricultural) and services, the United States has a comparative advantage in the production and exportation of seed. Data for seed exports by crop are shown in Table 4-1. Trade in seed is growing faster than use, indicating that seed has been an undertraded commodity. Because information on the capabilities and availability of improved seed is spreading to prospective users, new opportunities for trade are still to be found. As long as a significant proportion of potential users of improved seed are unaware of the products available, seed will be undertraded, and the potential for the growth in trade will exceed the growth in the use of seed. The data indicate that U.S. exports grew in volume terms by roughly 12 percent a year, doubling in six years between the mid-1970s and early 1980s. Imports grew more slowly, but still exceeded the growth in U.S. use of improved seed.

Table 4-1 shows that the largest categories (grasses and forage and vegetable seed) have been among the slowest growing. The strongest performers have been corn, soybeans and sugar beets, with the latter starting from zero as late as the 1976/77 seed season. (Seed seasons are from July 1 to June 30 of the next calendar year.) Soybean sales seem to be a huge and rapid success, reflecting positively on the effect of U.S. public breeding research carried out in the early and mid-1970s. The aggregate data show some surges and declines, but generally indicate strong growth over the period. A look at the trends of specific crops shows somewhat more volatility, which is part of the prevailing underlying pattern. Corn seed exports, for example, rose by 77 percent between 1975/76 and 1976/77, then fell 17 percent the next year. Soybean exports rose by 71 percent in 1981/82 then fell by 50 percent. "Other seeds" increased 50 percent in 1980/81, then declined sharply by almost 50 percent. The first two surges are explained by large purchases by single countries, Russia and Mexico, and the last is 60 percent accounted for by two countries, Japan and the Netherlands.

A final point to note is the surprisingly low rate of price inflation for seed exports. The percent increase from 1975/76 to 1982/83 was

TABLE 4-1. U.S. EXPORTS OF SEED WORLDWIDE
(Value in U.S. $ Thous.)

Types of Seed	1975/76	1976/77	1977/78	1978/79	1979/80	1980/81	1981/82	1982/83	1983/84
Grasses & forage	36,037	48,317	55,598	52,467	74,866	78,207	63,006	67,398	63,662
Vegetables	54,286	51,982	58,571	60,727	81,277	103,151	98,424	106,131	125,981
Flowers	2,709	2,792	3,379	3,126	3,090	4,324	4,857	5,308	7,484
Corn*	13,142	23,281	19,302	17,063	29,707	41,555	53,795	62,926	58,196
Sorghum	10,999	14,159	16,784	13,952	25,493	27,692	27,959	27,010	31,803
Soybeans	0	0	873	7,746	1,654	13,732	23,477	11,721	16,005
Sugar beets	0	0	1,841	642	1,683	1,811	2,149	3,345	4,098
Other	7,249	7,614	12,665	13,875	17,363	26,114	16,057	20,778	20,661
Total	124,422	148,145	169,013	169,598	235,133	296,586	289,724	304,617	327,890
Volume (in kilograms)	132,544	128,002	172,622	194,061	199,762	262,030	277,097	243,429	268,452
$/kg.	$0.94	$1.16	$0.98	$0.87	$1.18	$1.13	$1.28	$1.25	$1.22

*Sweet corn is included with vegetables.

Source: USDA.

33 percent, or an annual average of 4.2 percent in U.S. dollar terms. This is significantly less than the U.S. inflation rate and, even adjusting for the appreciation of the dollar over the period, it is less than inflation in most other countries. Moreover, when quality enhancements in terms of higher yields are included, the price increase per unit of output generated is almost certainly less than 4 percent per year. Thus, the initial evidence would suggest that traded goods tend to be price competitive, and imports can help keep prices down within local markets.

The destination of U.S. exports by country is shown in Tables 4-2 and 4-3 in two forms. Table 4-2 lists the leading importers by countries and regions; Table 4-3 ranks the top 12 importers of U.S. seed for various years over the last decade. Two points emerge from these tables to support the earlier conclusion that seed is an undertraded commodity relative to the potential size of the market: U.S. seed exports are highly concentrated in just six countries; and there is a great deal of volatility in purchases within crop categories. The volatility issue can be readily seen in the country data in Table 4-2. In 1976/77, the Soviet Union entered the U.S. seed market to buy $12 million worth of corn seed and within two years its purchases were insignificant. In 1980/81, Poland purchased almost $8 million of corn and forage seed, yet it was out of the market the next year. In 1981/82, Mexico imported almost $20 million worth of soybean seed, totaling 84 percent of U.S. soybean seed exports. The following year, Mexican purchases of soybeans fell to $4.3 million. In the case of Poland and the Soviet Union, the motivating factors were probably a combination of making up for shortfalls in domestic seed production (security) and the desire to substitute imports of seed for imports of grain (balance of payments). Corn, of course, is not an open-pollinated crop, so the potential benefits from introducing U.S. hybrids into the local genetic pool are not clear. Mexico, on the other hand, should derive a continuing benefit from the introduction of new improved soybean varieties into local agriculture. Security and balance of payments motivations may also have been at work, but upgrading the genetic base of the domestic soybean crop was probably a critical consideration. Political factors can contribute to volatility as well. In the late 1970s, Iran was a growing market for U.S. seed exports, but after 1979 political relations deteriorated to the point where exports fell to zero by 1980/81. Saudi Arabia started from a lower base, but has become a large importer of American seed — mainly vegetables — in the 1980s. Again, however, this market could become much less favorable if political problems were to develop.

International debt also contributes to market volatility, reducing the amount of foreign exchange available to many Latin American countries to buy imports. The dollar value of U.S. seed sales to South America grew at a compound annual rate of 37 percent between 1975–

TABLE 4-2. U.S. EXPORTS OF SEED TO COUNTRIES AND REGIONS (Value in U.S. $ Thous.)

	1980/81 Percent	1975/76	1976/77	1977/78	1978/79	1979/80	1980/81	1981/82	1982/83	1983/84
Canada		22,655	25,492	27,212	25,438	33,058	34,420	31,623	29,420	29,472
Mexico		19,515	16,187	20,651	20,294	28,576	45,818	60,765	46,323	72,674
North America*	29.2	45,737	45,962	52,801	49,737	68,195	86,494	99,511	82,464	109,366
Argentina		1,003	1,340	2,775	2,695	8,526	8,008	2,147	2,679	3,857
Venezuela		1,771	4,054	5,226	5,679	7,967	9,094	7,820	4,460	6,392
Brazil		3,294	4,750	6,153	6,188	5,424	5,798	3,767	5,166	5,107
South America*	10.9	8,242	12,958	17,672	19,866	29,235	32,358	21,384	20,521	25,573
United Kingdom		7,274	5,847	6,987	7,922	10,398	9,911	8,856	11,071	8,746
Netherlands		5,324	7,702	8,889	8,716	15,163	18,579	13,058	19,406	13,514
France		7,830	9,486	10,030	8,462	10,070	10,785	14,591	18,589	12,206
Germany		3,608	6,886	6,941	4,906	6,390	8,412	6,770	8,179	5,687
Italy		6,123	7,907	13,102	14,122	19,775	23,337	26,813	32,389	24,634
Greece		760	576	656	1,923	3,918	6,112	7,449	5,884	4,196
Spain		3,012	1,701	2,405	3,418	6,577	4,857	4,042	4,137	3,572
Sweden		1,746	1,498	2,119	2,051	3,536	3,592	3,806	3,242	4,276
Austria		198	405	549	506	1,014	1,690	591	3,628	1,041
Hungary		1,408	607	533	884	690	1,184	1,247	288	385
Romania		243	7	16	48	386	2,230	695	721	783
Poland		1,121	1,812	486	662	1,576	7,983	0	1,682	189
USSR		1,113	12,064	2,567	71	19	206	143	84	21
Europe*	36.2	43,496	61,315	60,036	58,813	85,218	107,338	96,902	119,947	90,037

TABLE 4-2 Continued

	1980/81 Percent	1975/76	1976/77	1977/78	1978/79	1979/80	1980/81	1981/82	1982/83	1983/84
Libya		427	784	586	358	1,147	1,644	1,570	2,406	3,390
Egypt		609	335	1,024	1,178	962	3,430	5,249	2,512	3,640
South Africa		3,549	1,636	1,498	1,315	1,880	3,233	5,453	6,510	6,914
Africa*	3.5	5,925	4,016	5,273	4,933	6,174	10,318	14,574	12,729	13,892
Iran		2,204	2,222	2,972	2,713	1,926	0	553	1,467	1,121
Japan		9,889	10,904	15,775	17,405	21,477	31,331	29,967	34,427	40,752
South Korea		105	221	694	401	258	3,441	962	2,655	4,088
Saudi Arabia		545	994	1,319	1,439	1,970	2,921	5,315	10,319	11,464
Asia*	17.4	18,015	19,265	27,094	29,605	36,674	51,638	48,948	61,305	81,570
Australia		2,702	3,452	5,363	5,737	8,145	7,194	7,867	6,278	6,193
Oceania*	2.8	3,047	4,607	6,137	6,645	9,603	8,438	8,627	7,627	7,453
Total	100.0	124,462	148,123	169,013	169,599	235,099	296,584	289,946	304,593	327,891

*Includes countries other than those shown.

Source: USDA.

TABLE 4-3. LEADING IMPORTERS OF U.S. SEED
(Value in U.S. $ Mill.)

1975-76		1976-77		1978-79		1980-81		1981-82		1982-83		1983-84	
Canada	22.7	Canada	25.5	Canada	25.4	Mexico	45.8	Mexico	60.8	Mexico	46.3	Mexico	72.7
Mexico	19.5	Mexico	16.2	Mexico	20.3	Canada	34.4	Canada	31.6	Japan	34.4	Japan	40.8
Japan	9.9	USSR	12.1	Japan	17.4	Japan	31.3	Japan	30.0	Italy	32.3	Canada	29.5
France	7.8	Japan	10.9	Italy	14.1	Italy	23.3	Italy	26.8	Canada	29.4	Italy	24.6
U.K.	7.3	France	9.5	Netherlands	8.7	Netherlands	18.6	France	13.1	Netherlands	19.4	Netherlands	13.5
Italy	6.1	Italy	7.9	France	8.5	France	10.8	Netherlands	14.6	France	18.6	France	12.2
Netherlands	5.3	Netherlands	7.7	U.K.	7.9	U.K.	9.9	U.K.	8.9	U.K.	11.1	Saudi Arabia	11.5
Germany	3.6	Germany	6.9	Brazil	6.2	Venezuela	9.1	Australia	7.9	Saudi Arabia	10.3	U.K.	8.7
South Africa	3.6	U.K.	5.8	Australia	5.7	Germany	8.4	Venezuela	7.8	Germany	8.2	South Africa	6.9
Brazil	3.3	Brazil	4.8	Venezuela	5.7	Argentina	8.0	Greece	7.4	South Africa	6.5	Venezuela	6.4
Spain	3.0	Venezuela	4.1	Germany	4.9	Poland	8.0	Germany	6.8	Australia	6.3	Australia	6.2
Australia	2.7	Australia	3.5	Spain	3.4	Australia	7.2	S. Arabia	5.3	Greece	5.9	Germany	5.7
Top 4	48.1%		43.7%		46.6%		45.4%		51.5%		46.7%		—
Top 6	59.9		55.4		55.7		55.4		61.0		59.2		—
Next 6	17.3		22.0		17.9		12.8		14.7		15.8		—

Source: Based on data in Table 4-2.

76 and 1979–80; in 1980–81, growth fell to 6.8 percent and exports have been declining since then. Financial stringency has evidently dictated that seed imports be postponed, but if this continues, the losses in agricultural production could begin to mount. Brazil and Argentina are agricultural exporters, while Mexico and Venezuela are importers, so the postponement of seed imports has probably begun to cost them foreign exchange in terms of reduced exports and/or increased imports. By contrast, exports to Europe grew more slowly in the 1970s, 18 percent annually in value terms, but have continued to increase in the 1980s with surges in 1980–81 and 1982–83. Growth of seed sales to Japan has also been steady and has continued to rise in the 1980s. Canada has traditionally been the top importer of U.S. seed, but has declined in importance in recent years. Table 4–4 lists U.S. volume exports to countries and regions and shows clearly that Canadian imports of U.S. seed have not grown appreciably since the mid-1970s. Canada shows signs of being a mature market in terms of seed imports, and future growth in trade will be in line with increased demand for seed. With regard to Canada, seed is not "undertraded" because most of the potential purchasers are using improved seed and are already fully aware of the products available from domestic and international suppliers.

Seed export markets may thus be divided into four broad types: centrally planned countries; less advanced developing countries; advanced developing countries; and the developed countries. Centrally planned countries are usually small purchasers, but intermittently they enter the world seed market in a large and unpredictable way. They are the most volatile markets. The less advanced developing nations have a pattern wherein they buy small amounts on a regular basis, with occasional surges in purchases. Their regular purchases are comparatively greater than the centrally planned countries and their surges are more modest. For both types of countries, the volatility of their purchases is more pronounced than their growth. Among the centrally planned economies, Hungary shows unusual stability in its purchases. Libya and Egypt are perhaps examples of the less advanced developing countries in terms of seed importing patterns. Egypt's political shift toward the United States in the 1970s no doubt helped increase seed exports, whereas Libyan imports held steady at lower levels in the early 1980s. The more advanced developing countries are at the stage of agricultural development where they need more improved seed and do not yet have the domestic capacity to supply their own needs. They are in a position to use imports to upgrade the genetic base of their seed stocks and improve their breeding programs, and this helps to explain the growth in their imports in the 1970s. However, foreign exchange problems following the second OPEC price rise in 1979–80 and the ensuing debt crises have, at least temporarily, cut off and reversed this upward trend.

TABLE 4-4. U.S. EXPORTS OF SEED TO COUNTRIES AND REGIONS
(Volume in Metric Tons)

	1975/76	1976/77	1977/78	1978/79	1979/80	1980/81	1981/82	1982/83	1983/84
Canada	33,957	26,829	25,128	24,303	29,201	29,444	27,983	29,011	27,140
Mexico	29,672	22,256	29,819	64,949	43,650	52,734	74,561	50,536	79,562
North America*	66,585	52,255	58,043	92,464	77,212	86,732	107,034	82,358	110,200
Argentina	358	612	2,380	2,571	4,637	6,261	691	1,298	1,271
Venezuela	1,322	4,750	7,576	5,935	7,972	8,532	7,076	3,663	5,572
Brazil	1,727	3,528	1,852	4,256	1,874	1,499	805	813	849
South America*	4,687	10,797	13,765	17,206	19,506	21,237	13,019	9,760	12,999
United Kingdom	7,186	4,158	7,635	11,918	12,756	12,090	9,938	14,561	14,026
Netherlands	4,442	5,039	6,995	9,506	10,049	24,124	10,621	21,561	9,386
France	8,208	7,684	8,296	6,668	6,954	7,513	10,971	9,040	7,302
Germany	3,134	6,726	6,626	6,084	7,702	12,104	6,691	10,137	7,558
Italy	5,036	5,555	9,931	10,381	13,355	15,172	15,988	22,340	14,608
Greece	176	118	305	1,378	2,173	3,384	3,666	2,318	1,714
Spain	2,527	1,189	1,760	2,532	4,867	4,001	2,372	2,100	2,005
Sweden	1,671	1,418	1,479	2,155	3,415	3,323	3,923	2,566	3,850
Austria	230	326	355	243	622	1,035	501	1,915	274
Hungary	957	496	769	677	799	789	515	312	123
Romania	203	3	15	62	294	1,231	167	164	174
Poland	614	725	236	486	950	1,677	0	993	16
USSR	1,972	9,967	25,461	14	1	62	27	3	1
Europe*	38,752	46,606	72,533	56,215	67,943	94,440	69,988	96,192	71,669

TABLE 4-4 Continued

	1975/76	1976/77	1977/78	1978/79	1979/80	1980/81	1981/82	1982/83	1983/84
Libya	109	107	276	194	149	327	124	104	257
Egypt	51	44	633	1,426	475	1,150	2,455	598	1,450
South Africa	2,586	1,150	832	529	1,191	1,240	1,790 —	5,089	7,117
Africa*	3,351	2,442	2,445	4,790	4,201	3,296	5,326	6,484	10,159
Iran	602	487	887	478	204	0	33	155	55
Japan	15,117	11,544	17,534	16,222	22,685	40,091	22,800	28,092	30,806
South Korea	123	187	658	408	310	5,317	1,009	3,020	4,642
Saudi Arabia	108	190	240	331	400	1,147	2,595	9,346	14,508
Asia*	17,446	14,030	23,195	20,682	27,361	52,831	28,326	44,547	59,116
Australia	1,607	1,843	2,338	2,337	2,970	2,979	3,187	3,570	3,859
Oceania*	1,722	2,342	2,641	2,704	3,539	3,653	3,385	4,062	4,309
Total	132,543	128,472	172,622	194,061	199,762	262,189	227,078	243,403	268,452

*Includes countries other than those shown.

Source: USDA.

The fourth type of country is the advanced market economy, which is more stable than any of the other types, but is innately slower growing than the Latin American countries. Canada, Australia, Europe, and Japan are more mature markets with comparatively modern agricultural sectors. Most farmers in these countries have been using improved seed for many years. Improved seed is not as undertraded in these markets as in the rapidly developing agricultural sectors in Latin America. Volatility in demand for imports, however, is present even in the developed economies. A closer look at the kinds of seed exported by the United States to the European Economic Community, shown in Table 4-5, reveals some underlying volatility in grasses and forage, corn and other seed. On balance, though, this market, together with the Canadian and Japanese markets, must be regarded as comparatively stable. They are not given to the large-scale cumulative miscalculations that periodically occur in centrally planned agricultural systems. They also do not have foreign exchange scarcities. Perhaps equally important, they are relatively free and open economies when it comes to importing seed. The matter of trade barriers will be discussed in detail in Chapter 8, and while such barriers are a fact of life in all countries, they are more predictable and more surmountable in Canada, Japan and the EC countries. The casual connection between openness to trade and stability probably runs in both directions. Openness is possible and acceptable because the agricultural sector is relatively mature, and demand for improved seed is stable. On the other hand, the openness is part of a market-oriented strategy that contributes to stability in both prices and volumes of seed. Diversity of suppliers helps avoid miscalculations that have the pervasive consequences seen periodically in centrally planned systems. Imports add to the potential diversity of supply, and if weather or genetic vulnerability should create an increased need for external supplies of seed, a country with established international ties can fill its needs with a minimum of difficulty and interruption. The same approach also fosters competition, as described above, and it is generally the case that countries with relatively free and open seed markets also tend to have relative stability of supply, price competition and steady improvements in seed quality.

American seed exports tend to be concentrated with regard to the country of their destination (note that some of the exports to the Netherlands are directed elsewhere after arriving in Dutch ports). Within the top dozen importers of U.S. seed (see Table 4-3), concentration ratios fell slightly in the late 1970s, but seem to be rising in the early 1980s. Since 1978/79, the top countries have remained the same — Canada, Mexico, Japan, Italy, France, and the Netherlands. Together they usually purchase 55 to 60 percent of all U.S. seed exports. The United Kingdom, Germany, Australia, and Venezuela have been con-

TABLE 4-5. U.S. EXPORTS OF SEED TO THE EUROPEAN ECONOMIC COMMUNITY
(Value in U.S. $ Thous.)

Types of Seed	1975/76	1976/77	1977/78	1978/79	1979/80	1980/81	1981/82	1982/83	1983/84
Grasses & forage	8,170	15,591	17,320	13,125	20,967	21,533	14,419	18,226	14,389
Vegetables	17,114	16,171	18,484	17,617	24,075	28,231	25,631	31,090	30,113
Flowers	1,344	1,417	1,557	1,493	1,369	1,743	2,311	3,121	4,045
Corn*	3,981	3,585	5,789	8,288	10,533	12,053	25,536	32,808	16,863
Sorghum	611	1,076	1,751	1,257	1,763	1,581	2,323	2,010	1,688
Soybeans	0	0	24	585	299	283	1,469	3,557	1,057
Sugar beets	0	0	531	382	575	583	267	808	1,172
Other	1,526	2,639	2,994	4,071	5,278	8,732	4,619	7,727	5,996
Total	32,746	40,479	48,450	46,818	64,859	74,739	76,575	99,347	75,323
Volume (in kilograms)	29,517	30,300	40,918	47,156	53,154	73,121	56,501	83,188	59,837
$/kg.	1.11	1.34	1.18	0.99	1.22	1.02	1.32	1.19	1.26

*Sweet corn is included with vegetables.

Source: USDA.

sistently among the second six in this period. The share of this second group seems to have fallen from 17 or more percent of total exports in the 1970s to roughly 15 percent in the 1980s. Approximately 75 percent of U.S. seed exports go to the top dozen countries in a typical year, compared with 62 percent of total agricultural exports.[6] Whereas 45 to 50 percent of U.S. seed exports are directed to just four countries, the comparable figure is 38 percent for all agricultural exports.[7] Concentration of agricultural exports is also declining over time, yet concentration in seed exports seems to be rising. It will be interesting to see if the concentration ratios of seed exports continue to rise in the years ahead. If the South American countries return to the international market, U.S. exports to this region should increase and overall concentration levels should then decline. This is a fast growing market for improved seed, while Europe and Japan are slower growing. During the 1980s, growth in these two more mature markets could begin to level off as has been the case in seed exports to Canada. Trends along these lines would reduce concentration, but possibly increase volatility in U.S. exports of improved seed.

In terms of the EC's trade balance for wheat, vegetable and corn seed, Table 4–6 indicates that Europe is running a small surplus as a seed trader for wheat and vegetable seed. Note that some of the exports are to countries within the EC so the same transaction will appear as both an export and an import in this table if it is intra-EC trade. This explains why exports and imports tend to surge in the same years. Also note that the years are calendar years, not seed seasons.

Soft wheat is the more important wheat crop in Europe and, as in most locales, the farmers tend to save seed and replant from the previous year's harvest. Therefore, only about 40 percent of seed needs are purchased from seed companies in a typical year. The first important reality to be aware of is that international trade is a very small part of overall production and use of seed. Based on data for area sown and typical use per hectare, wheat seed consumption in France was in the range of 600 to 640 million kg. in the period 1975 to 1982. And, as France has about one-third of the total wheat area of the EC, it is clear that imports were a very small part of total wheat sown, in the range of 1 percent. Relating imports to wheat seed purchased, the ratio rises to perhaps 2.5 percent, still a small amount. Therefore, trade is largely a residual factor making up for shortfalls that countries may face periodically. The peaks in soft wheat seed imports shown in 1975, 1979 and, to a lesser degree, in 1982, are indicative of this motivation for imports. Hard wheat is not as important a crop, and the most interesting aspect of these trade flows is the surge in exports that occurred in 1977 and 1978. These are almost entirely exports from France to Algeria and indicate that evidence for volatility in trade flows is not limited to U.S. trade data.

TABLE 4-6. TRADE IN SOFT WHEAT SEED, HARD WHEAT SEED AND VEGETABLE SEED, EC COUNTRIES, 1975-82 (MIll. kg./Mill. ERE)

| | Soft Wheat Seed | | | | Hard Wheat Seed | | | | Vegetable Seed | | | |
| | Exports | | Imports | | Exports | | Imports | | Exports | | Imports | |
	Volume	Value	Volume	Value	Volume	Value	Volume	Value	Volume	Value	Volume	Value
1975	16.7	3.7	26.2	4.7	1.7	0.5	0.7	0.1	10.1	44.8	12.8	40.5
1976	9.7	2.6	3.8	1.0	0.5	0.1	1.2	0.3	11.4	55.9	13.2	50.5
1977	8.1	2.2	8.9	2.1	11.8	5.1	1.5	0.3	12.7	66.6	14.5	60.3
1978	11.6	3.6	7.1	2.1	6.7	3.0	0.2	0.1	12.0	70.2	15.1	61.3
1979	21.7	6.7	18.3	5.6	2.2	0.8	0.2	0.1	14.8	88.1	15.8	67.5
1980	9.6	3.3	9.8	2.9	1.3	0.4	0.2	0.1	12.5	90.6	15.5	70.2
1981	27.8	8.3	11.5	3.8	1.1	0.4	1.5	0.5	14.4	104.8	14.5	81.5
1982	25.9	7.4	17.3	5.9	1.7	0.7	3.2	1.2	13.9	121.2	14.7	92.4

Source: Eurostat, various years.

EC trade in vegetable seed, as with the United States, is a much higher proportion of total seed use, comprising roughly 30 percent. The vegetable seed trade is not a residual, but a part of the core of supply and demand, and therefore trade flows are more stable over time. The volumes are no larger than traded wheat seed, but the values are much higher because of the higher cost of vegetable seed production and the greater value of the crop produced per unit of seed. The import market would also appear to be quite mature in that the volume of trade has not grown significantly since the mid-1970s and after peaking in 1979 has declined.

Within the EC, France is unquestionably the leading agricultural country, and this leadership extends to seeds. France is the largest producer, user and exporter of improved seed in Europe. According to the National Interprofessional Group for Seeds and Plants (GNIS), the leading national organization dealing with the seed industry, French seed exports in 1981 were about $185 million, compared with U.S. seed exports of about $290 million. French seed exports increased their share of world trade from 8.5 percent in 1970 to 11.5 percent in 1980.[8] Data on the breakdown of French seed production, presented in Table 4-7, show the types of seed exported, the numbers of species and varieties available, and the total acreage utilized for accredited seed cultivation. The importance of small grain cereals, mainly wheat and barley, for domestic use is evident from the acreage in seed. In value terms, corn is the leading seed export, with vegetables and flowers in second place. It is clear from this array of species and varieties that France produces a broad range of seed and has the potential to continue expanding.

The importers of French seed tend to be located in close proximity, with the other EC countries accounting for 50 percent of total exports in 1981. Other Western European countries purchased 19 percent, while North Africa accounted for 11 percent, Eastern Europe 8 percent, and the Middle East 7 percent.[9] Over the years, France has built up an integrated seed industry with all the characteristics of a seed exporter discussed in the previous section. Knowledge and experience have accumulated on both the research and production sides to make France the second leading seed exporter in the world. There is at least one cloud on the horizon, however, and that is the effect that the EC's Common Agricultural Policy subsidies on agricultural output are having on seed production. Very few seed crops are subsidized under the CAP (some forage and oil crop seed), so seed producers must compensate farmers for growing unsubsidized seed, rather than subsidized crops. This pushes up seed producers' costs and is beginning to have an effect on the price competitiveness of French and other EC seed producers. The next chapter will deal with this problem as it applies to EC corn seed production.

TABLE 4-7. FRENCH SEED PRODUCTION, 1981

	Species	Varieties	Hectares**
Small grain cereals	7	252	188,030
Corn	1	157	39,400
Sorghum	1	3	n.a.
Forage crops	22	233	56,100
Vegetables*	45	870	10,000
Potatoes	1	76	12,890
Beets	1	104	4,135
Fiber crops	2	14	3,620
Oil and protein crops	8	65	12,070
Total	88	1,774	326,245

*Includes flower seed.
**Acreage utilized for accredited seed cultivation.
Source: *French Seeds* (Paris: GNIS, 1982).

This review of international trade in seed in Western Europe generally supports the findings of the analysis of U.S. seed exports. While data are often unavailable or difficult to collect in a systematic form, it is still possible to draw several conclusions with reasonable confidence. It appears that the countries of the EC are becoming mature economies in terms of imports of improved seed. Just as occurred in the late 1970s with Canada, these markets will soon have exhausted their untapped potential for absorbing additional volumes of improved seed. Growth in export sales will then have to come by wrestling market shares from existing varieties rather than by moving into new and undeveloped areas. Japan is probably not as far along in this process, but by the end of the 1980s could be reaching this stage. On the export side, the EC countries, particularly France, have the capability to expand if they can keep production costs down and remain price competitive. The import markets with the greatest potential for growth are in Latin America, Asia and perhaps the Middle East, but the foreign exchange shortages currently facing many of them will inhibit their imports of seed for many years to come. These countries are discussed in more detail in Chapter 7. For both France and the United States, a high proportion of seed exported is concentrated in several markets, most of which are at or approaching maturity in terms of import growth. U.S. exports seem more diversified by region with markets established in Asia and Latin America, in addition to Europe and North America. France seems more focused on relatively mature markets than the United States and thus may face more stability of demand, but reduced potential for growth. Volatility is a problem for all exporters and will probably increase in the years ahead as developing countries become more important purchasers.

FOOTNOTES, CHAPTER 4

1. For members of the European Economic Community, there are important consequences for the subsidies paid to agricultural producers under the Common Agricultural Policy discussed in the next chapter.
2. For more on the theory of comparative advantage, see a textbook in international economics, e.g., *International Economics*, Charles P. Kindleberger and Peter W. Lindert (Homewood, Ill.: Irwin, 1978).
3. See Raymond Vernon, *Quarterly Journal of Economics* (May 1966).
4. For more on these issues with regard to Mexico, see Louis W. Goodman, with Arthur L. Domike and Charles Sands, *The Improved Seed Industry: Issues and Options for Mexico* (Washington, D.C.: Center for International Technical Cooperation, 1982).
5. Proportions derived by comparing exports to U.S. market for seed as presented in Table 3-2 earlier.
6. Average of years 1970 and 1980; see Timothy Josling, *Problems and Prospects of U.S. Agriculture in World Markets* (Washington, D.C.: NPA, 1981), p. 10.
7. Ibid.
8. *French Seeds* (Paris: GNIS, 1982), p. 28.
9. Ibid., p. 29.

Hybrid Corn Seed in the European Economic Community

5

This chapter is a case study of recent developments in the international trade of hybrid corn seed in Western Europe. Detailed analysis of a particular crop and market will help to provide a deeper understanding of current issues and emerging problems in the international trade of improved seed among developed countries. Many of the general propositions made in the previous chapters will be clearer when viewed as a concrete case study. The discussion will begin with recent history of the cultivation of corn in Western Europe. This will be followed by a description of the current situation, particularly of the distortions in the seed market created by the Common Agricultural Policy. Genetic diversity and the lack thereof in some countries will be mentioned, as well as the recent increase in imports from Eastern Europe. The role of the General Agreement on Tariffs and Trade (GATT) will also be covered. The discussion does not include Spain or Portugal, which have only recently become members of the EC.

BACKGROUND ON HYBRID CORN SEED IN EUROPE

During the Age of Discovery, corn was brought to Europe by the early explorers of North and South America. While continuously cultivated since the 1500s in various countries around the Mediterranean, including Turkey, corn was never a leading crop in their agricultural systems. Over the centuries, however, local varieties of corn became well adapted to their new homes. Occasionally, varieties were exchanged between countries, with the expected benefits from increased genetic diversity, but there was never a true breeding program and hybrids were not introduced on a significant scale until after World War II. Before 1949, the corn varieties grown in Europe were highly susceptible to lodging problems and low yields. Historical data for the period 1948–52 reveal that the yield of corn in metric tons (m.t.) per hectare (ha) was 0.36 for France and 0.55 for Spain,[1] compared with roughly 2.0 for the United States. The areas planted were very small, comprising only about 3 percent of cultivated land in France and 4 percent in Spain. The only other producer of any significance at this time was Italy.

The recent history of corn seed production in Western Europe is the story of developments in central and northern France. Southern France, Spain, Italy, and Greece face different ecosystems and are able to utilize American hybrids directly. Few if any areas in all other Western European countries are warm enough to utilize direct imports from the United States as commercial corn seed. The great preponderance of corn cultivation north of the Alps requires corn varieties with an earlier maturity than typical U.S. varieties. In the late 1940s, American hybrids, principally from the University of Wisconsin, were introduced into France. These varieties were suitable for cultivation mainly in the warmer regions of France and were only partially successful, but they began the process of introducing American varieties. In the early 1950s, these varieties were the principal commercial varieties in France.

At the same time, the French National Institute for Research in Agronomy was establishing a serious breeding program, beginning by collecting all available French corn varieties. By the middle 1950s, INRA had developed two good parent lines for hybrid corn. One or the other of these have provided half the parentage of most commercial varieties of hybrid corn since developed in France. Crosses of one of these two locally adapted inbreds with an American inbred to make a single cross is basic to the hybrid corn seed industry in France and northern Europe. Perhaps most important, INRA's research helped make possible the first early maturing varieties of corn. Introduced in 1958, INRA 258 had an FAO number of 250 certifying that this variety could be cultivated in much colder environments than traditional corn varieties. American corn varieties are typically rated at FAO 400 and up, requiring proportionately more heat units to reach maturity. Today, some of the earliest French varieties are rated at less than 200, but for its time INRA 258 was a significant breakthrough, and it began the trend that pushed corn cultivation further and further north in Europe. In the late 1950s and early 1960s, INRA expanded its share of the total French market to 40 percent, despite the fact that it sold only through agronomists and co-ops and not through an extensive network of dealers.

Agricultural cooperatives are very important in France, providing purchasing, marketing, seed production, and many other services for their members. They are strong institutions with a great deal of influence on the politics and the people within the departments (regions) that contain their members. In some ways, they are the modern, democratic versions of the old manoral system. The leaders of the cooperatives are loyal to the local people and are expected to advance the interests of their members at regional and national levels. One by one in the late 1950s and 1960s, the cooperatives began to move into the seed business. In the late 1950s, the Co-op du Pau, working with the Funk Seed Company of the United States, began to develop private commercial hybrids by crossing INRA parent lines with Funk lines. Their efforts were successful, and by the mid-1960s these hybrids had become competitive, with respectable shares of the total French market. By this time, other co-ops had launched breeding programs in conjunction with many of the leading U.S corn seed companies, including Cargill, DeKalb, PAG, Northrup-King, and Pioneer Hi-Bred. For the most part, research was conducted jointly, using various inbred lines provided by the U.S. partner crossed with the proven INRA lines. Production was principally in the hands of the French partner, with the Americans investing some capital and the initial technical expertise. The French government encouraged each of these producers to set up their own production operations as part of the process of entering the national market. By so doing, the seed companies have occasionally found themselves with excess capacity in receiving, storing, sorting, treating, and bagging corn seed. During growth periods, this has not been a problem, but when growth slackens, any excess capacity has added to the cost problems facing the French hybrid corn seed industry.

Throughout the quarter century after 1950, acreage devoted to corn continued to expand and yields rose steadily, reflecting the results of breeding research. Table 5–1 provides data on corn acreage and production in France, showing the results of the breeding successes of INRA in the 1950s and private breeders in the 1960s and 1970s. In the 1970s, corn was introduced in the northern limits of the French zone of corn cultivation. By 1974, apparently too much land in colder regions had been planted, as yields of grain corn were down. Thereafter, marginal land shifted into silage corn production, and yields of grain corn again rose. Silage, or green, corn is harvested before the plant reaches maturity, i.e., before the corn kernels are fully developed on the ears. The whole plant is harvested, ensiled and used as fodder, principally for cattle. While silage corn can be grown in regions where it is too cold to grow grain corn, the seed multiplication for silage corn must take place in warmer regions so that the kernels can reach full maturity and be harvested for use as seed by farmers farther north. For this reason, all the important seed producing cooperatives in France are in the warmer regions.

Returning to the discussion on French breeding, it is interesting that the biggest success story of the 1970s was not that of a cooperative matched with an American partner; rather, it was Limagrain, a cooperative in the center of France near the city of Clermont-Ferrand, that developed a new improved early variety called LG–11. Previously, Limagrain had been more active in small grain seed, but in the 1970s it shifted into hybrid corn seed production. The new variety LG–11, inscribed in 1970, was a triple cross based on work done by INRA. It is very hard to multiply this variety and maintain quality control, but Limagrain mastered the technology. It set up rigid controls that, over

TABLE 5-1. BASIC DATA ON CORN ACREAGE AND PRODUCTION (GRAIN AND SILAGE), FRANCE, 1948–82

	Corn Acreage and Production (Grain)						
	1948–52	1961–65	1968–69	1974–75	1978–79	1981–82
Area (1,000 ha)	332	914	1,105	1,933	1,875	1,595
Production (1,000 m.t.)	452	2,760	5,565	8,447	9,965	9,355
Yield (m.t. per ha)	1.36	3.02	5.04	4.37	5.31	5.87

	Corn Acreage and Production (Silage)						
Area (1,000 ha)	n.a.	n.a.	n.a.	845	1,095	1,255
Production (1,000 m.t.)	n.a.	n.a.	n.a.	34,470	46,581	50,263
Yield (m.t. per ha)	n.a.	n.a.	n.a.	40.79	42.54	40.05

Sources: Edgar Rihm (Semillas Pioneer, S.A., Seville, Spain), 1948–69; Eurostat, 1974–82.

time, became characteristic of the organization. The company's success was based on the superiority of LG-11, rigorous control of multiplication, adequate markets both in France and abroad, and modern, well designed production facilities. Limagrain quickly dominated the market for early grain corn production, and much of the expansion in corn acreage occurring in the 1970s was due to the strength of LG-11. After the mid-1970s, this variety was used as silage corn in the most northern areas and as grain corn in the semi-early regions. LG-11 did not perform well in the southern regions of France, but was never intended for those regions.

The French corn seed market can be divided into five parts based on the maturity index of the varieties that will grow best; the five maturity groups are listed in Table 5-2 with relevant characteristics. The number of available varieties has increased over time as new ones are introduced and old ones retain a niche in the market. The proportions of market share shift somewhat because many regions can use varieties in either of two groups. The keenest competition is in the early to semi-early groups in regions able to utilize varieties with maturity indexes in the range of 240 to 300. The strength of Limagrain was almost entirely based on one very good variety, which, by the late 1970s, dominated all the early groupings of the French market. Moreover, LG-11 was eminently exportable to other EC countries, principally Germany which required early varieties. In 1977–78, Limagrain sales peaked at roughly 40,000 tons of corn seed, half in France and half in export markets.

During the 1970s, Limagrain's successes were significant, but were predicated more on production and cost controls than a sound long-term research program. Limagrain was seen as a model agro-industry organization by the French government, at that time headed by Giscard d'Estaing, who was also from the Clermont-Ferrand region. The French government encouraged Limagrain to diversify, and it eventually used its seed profits to absorb a number of weak companies in unrelated

TABLE 5-2. FRENCH MATURITY GROUPINGS OF CORN SEED, 1983

Maturity Group	Varieties	Approximate Proportion of Seed Market	Maturity Index
Very early (silage)	43	18%	Less than 220
Early	52	34	180 to 260
Semi-early	46	32	240 to 360
Semi-late	29	8	330 to 470
Late	50	9	Above 470

Source: *Varietes de Mais: 1983*, Association Generale des Producteurs de Mais.

areas of agriculture. With its background and strength in controlling cost and production processes, Limagrain was not well suited to evolve into a conglomerate with operations in diverse activities, and the attempt at diversification has thus far been a drain. However, Limagrain has a strong financial base, has performed fairly well with marketing,[2] and in recent years has put more resources into its breeding research programs.

In the meantime, other seed companies have been introducing new improved varieties and taking sales away from Limagrain. The two main competitors are RAGT, a cooperative associated with DeKalb, and France-Mais, associated with Pioneer. RAGT has its strength in northwest France and its base in the south central region around Rodez. RAGT produces good early and very early varieties and is a significant exporter to West Germany. France-Mais is strongest in the early varieties and, therefore, competed with Limagrain in somewhat different areas than RAGT. France-Mais in particular has been taking market share away from LG–11 and by the mid-1980s has come to dominate the middle part of the French market—that is, the maturity range of 240–300 that is the heart of the grain corn market. In the late 1970s, France-Mais supplied 5 to 10 percent of the French market and was ranked third or fourth behind Limagrain and RAGT, and even with the Co-op du Pau. By 1983, France-Mais supplied approximately 40 percent of the French corn seed market, including as much as 75 percent of the middle range. Limagrain still leads the very early market, with RAGT also ahead of France-Mais in these varieties. However, throughout the middle and late maturities, France-Mais has grown to supply half or more of the French corn seed demand.

The basis for the rise of France-Mais has been an expanded breeding research program in the 1970s, new varieties with significant yield increases introduced in the early 1980s, and successful marketing activities. Some have criticized France-Mais in its marketing approach in the south of France. There are more small landholders in the south, and it is felt that a large network of small dealers is needed. France-Mais has been working through large co-ops, yet seems to be effective without the personal relationship that is traditional in the south of France.

There is no doubt that the breeding program Pioneer and France-Mais agreed to set up in 1972 has been the foundation of their successes in the 1980s. In the 1960s, France-Mais had a research program and received support from Pioneer, similar to the arrangements of several other French producers working with American firms. In the early 1970s, France-Mais and Pioneer reorganized their research program, set up a large modern research station in the Loire Valley, and agreed that Pioneer would be responsible for research and France-Mais for production and distribution. The breeding lines from the old research program were inherited by the new program and augmented by the introduction of capital, equipment and, in the beginning, breeding assistance from the United States. The results of these efforts can be seen in Table 5-3, which shows the date and maturity type of new varieties inscribed or registered in the official French catalogue of approved species and varieties. For each of the six organizations shown, the dates of inscription of the varieties available in 1983 are indicated; older inscribed varieties that have been dropped from the organizations' lists of commercially available varieties are not included. Inscription is the final step a breeder must surmount before a new seed variety can be marketed. It is the means by which the work and contribution of the breeder are officially recognized and sanctioned and requires that a variety must be unique (unlike existing varieties), uniform and stable. The variety must also pass field trials showing that it makes a contribution in terms of agricultural or user value.[3] For France-Mais/Pioneer, the data show a thoroughgoing renewal of their varieties beginning in 1978 and continuing in the early 1980s. The number of early varieties aimed at the largest part of the French market are shown in parentheses. Only one variety remains from before 1978 among the available varieties listed by France-Mais in 1983. As a percentage of France-Mais' available varieties, 94 percent were inscribed in 1978 or later and 61 percent in 1980 or later. The figures are similar for RAGT (87 and 65 percent) and Co-op du Pau (96 and 63 percent).

Gravadour/Cargill has 65 percent of its currently available varieties inscribed in 1981 or later, but none in 1978 to 1980. Limagrain and INRA have an even smaller proportion of recently introduced varieties. For Limagrain, only 67 percent are 1978 or later and only 44 percent are post-1980; for INRA, the ratios are 46 percent and 35 percent. The

TABLE 5-3. DATE OF INSCRIPTION OF AVAILABLE HYBRID CORN VARIETIES, TOTAL (EARLY)

	Limagrain	France-Mais	RAGT	Co-op du Pau	Cargill	INRA
1970	1 (1)	0	2 (2)	0	0	6 (4)
1971/74	1 (1)	0	0	0	3 (3)	4 (3)
1975	2 (2)	1 (0)	1 (0)	0	0	2 (0)
1976	0	0	0	0	1 (0)	2 (1)
1977	2 (2)	0	0	1 (1)	1 (1)	0
1978	3 (3)	5 (3)	2 (2)	4 (3)	0	1 (0)
1979	1 (0)	1 (0)	4 (3)	4 (0)	0	2 (0)
1980	0	3 (3)	6 (3)	4 (4)	0	4 (3)
1981	2 (0)	3 (0)	2 (1)	3 (2)	4 (2)	2 (1)
1982	2 (2)	2 (2)	2 (1)	1 (1)	4 (2)	2 (1)
1983	4 (3)	3 (2)	5 (5)	6 (6)	1 (1)	1 (1)

Source: French Catalogue of Inscribed Corn Varieties, as shown in *Varietes de Mais: 1983*.

INRA data may not be strictly comparable because INRA may be maintaining older varieties as a public service even though they are not in heavy demand. Therefore, most of Limagrain's competition renewed its varieties for the 1980s through successful internal research programs, while the market leader peaked in 1977–78 and did not introduce a single new early variety in the 1979–81 period. During this time, Limagrain's research programs were being reorganized and there was substantial turnover in key personnel. By 1980, the program had been reestablished as SIGA Limagrain Services and began to turn out new varieties, both early and late. Only time and competition in the market place will tell how successful the new varieties introduced in the 1980s will be, but the research efforts of France-Mais/Pioneer seem to be showing the best initial results. RAGT, Co-op du Pau and Cargill do not appear to have had serious deterioration in their market shares, while Limagrain's failure to make timely introductions of improved varieties in the late 1970s has reduced its sales levels noticeably.

Probably the most important problems currently facing the French corn seed industry are the high production costs and the growing reluctance of farmers to multiply seed at competitive cost levels. A large part of this issue is attributable to the Common Agricultural Policy, which subsidizes the price of grain corn but does not subsidize corn seed production. Typically, farmers contract with corn seed companies to multiply seed and are paid a multiple of the market price for grain corn as compensation for additional work and care in planting, cultivating and harvesting the seed plots.

Corn seed is planted in a pre-set ratio of male and female rows, and only the female rows that produce the hybrid seed are harvested. This means that only part of the corn actually planted and grown can be harvested and used for seed, whereas all the grain corn crop can be sold by the farmer. Thus, the farmer has to be compensated for reduced salable output per acre in addition to extra work, and experience has shown that it takes a multiple of five to six to attract a sufficient number of reliable farmers. The situation boils down to some straightforward propositions. Corn seed producers in the EC must pay farmers a multiple of a subsidized grain corn price, while corn seed is not subsidized. More and more EC corn seed producers are having to compete internationally with producers that do not face subsidized grain corn prices and thus have lower cost structures. Since the early 1980s, competition from U.S. production has been mitigated by the high value of the U.S. dollar relative to the French franc. This helped up until 1986, but still did not make EC corn seed production truly competitive. And, now that significant exchange rate realignments have occurred, the French and other EC corn seed producers could find themselves in a very difficult situation vis à vis trade with the United States in those varieties that adapt directly.

Problems were already surfacing in 1983 regarding trade with low cost Eastern European corn producers, most notably Hungary. French imports of seed produced in the East were still small, only about 5 percent of total consumption in 1983, but they caused considerable concern because of their price competitiveness. A general idea of the situation can be seen by comparing the cost estimates for the early 1980s presented in Table 5-4, which shows average French production costs and the average offer price of imports at the French border. The data make it clear that France is a high cost producer in comparison with Eastern Europe and would have a hard time competing with U.S. corn seed producers if the franc/dollar exchange rate were to stabilize in the region of 5 or 6 to 1, rather than 7 or 8 to 1 in the early 1980s. The issue of Eastern European competition will be explored in more detail after the discussion on Germany and Italy, the two other large markets for hybrid corn seed in the EC.

Developments in other European countries will be reviewed now. Beginning with Germany, the first important difference from France is that the climate is cooler and thus the early and very early varieties are required. In terms of corn cultivation, Germany is more homogeneous than France; in the north the altitude is low, and in the south the land rises. As a result, there is about the same temperature level throughout most of the country and similar maturities of corn apply widely. Between 70 and 80 percent of the German corn crop requires varieties with maturities in the 180 to 220 range.[4] Approximately 20 percent is warmer than this and can use varieties up to 300, while about 10 percent is colder and requires varieties in the 160 to 180 range. Almost all German production is silage corn, which, as noted above, is cut while green and used for fodder. Because the climate is too cold

TABLE 5-4. COMPARISON OF ESTIMATED COSTS, HYBRID CORN SEED FOR FRANCE AND EXPORTERS TO FRANCE (ECU per 100 kg.)

	Single Cross		Double Cross		Three-Way Cross	
	1981/82	1982/83	1981/82	1982/83	1981/82	1982/83
France	210	240	112	125	135	148
U.S.	190	195	85	150	140	148
Hungary	150	170	75	65	80	70
Romania	140	160	62	52	85	72
Yugoslavia	—	140	72	70	115	86
Austria	—	170	70	65	93	81
Canada	150	190	85	153	145	—

Sources: Eurostat and industry estimates.

TABLE 5-5. EC AREA DEDICATED TO CORN CULTIVATION (1000 ha)

	1978		1980		1982		Percent 1982	
	Grain	Silage	Grain	Silage	Grain	Silage	Grain	Silage
Germany	113	580	117	697	160	779	2.9	14.0
France	1,802	1,044	1,757	1,158	1,617	1,271	29.1	22.8
Italy	842	218	874	453	1,011	460	18.2	8.3
Netherlands	1	118	1	137	—	147	—	2.6
Belgium	6	83	6	90	7	98	0.1	1.7
Luxembourg	—	5	—	—	—	—	—	—
United Kingdom	1	26	—	22	—	16	—	0.3
Ireland	—	—	—	—	—	—	—	—
Denmark	—	—	—	—	—	—	—	—
Total (EC 9)	2,765	2,074	2,755	2,557	2,795	2,771	50.3	49.7

Source: Eurostat.

for corn to reach full maturity and produce fully developed kernels, Germany can produce only a very small part of its corn seed and is the great import market in Northern Europe. Table 5-5 presents data on the area of cultivation of both grain and silage corn for nine of the countries in the EC. This indicates the distribution of the corn seed market between countries and uses of the corn grown. The clearest trend over recent years has been the expansion of the cultivation of silage corn. In France, this has been at the expense of acreage in grain corn, but elsewhere it has been in addition to grain corn. By 1982–83, the area given over to the cultivation of silage corn was almost equal the area in grain corn. The only other notable time trends are the decline in grain corn acreage in France and the roughly equal rise in Italy. France is clearly the dominant country in the EC corn picture. Of total French acreage, 56 percent is in grain corn. Germany, by contrast, is about one-third the size of France in area of cultivation and has almost 83 percent in silage corn. Italy, approximately half the size of France, has about 70 percent in grain corn. The Netherlands and Belgium together comprise almost all the remainder and are predominantly silage corn producers. The rest of the discussion here will center on the three larger countries. The Netherlands and Belgium have agricultural systems, interests and policies basically similar to Germany on matters dealing with hybrid corn seed.

TRADE OF HYBRID CORN SEED IN EUROPE

A comparison of domestic hybrid corn production, together with imports and exports, is shown for the three leading EC countries in Table 5-6. Results are presented for the seed seasons of 1980/81 and 1981/82 and verify the general trends discussed above. France is the largest producer; exports take up about 40 percent of total production and imports are slightly more than 3 percent. For France, the most important external market is Germany, which purchased about 16 percent of total production and about 55 percent of French exports. Germany was highly dependent on imports supplying 36 percent of its apparent consumption from extra-EC sources, about 55 percent from France and other EC countries, and only about 10 percent from domestic production. Italy is also a net importer, but produces a much larger share of its consumption domestically. Moreover, Italy's imports came from outside the EC, principally the United States.

Germany is a classic importer with most of the importer's motivations described in Chapter 4. This includes security of seed supply with appropriately early maturity, diversity of supply, price competition, and access to high quality new varieties from various breeders. Germany imports because of climate, and a large share of the country's needs are met by German seed firms who contract out seed multiplica-

**TABLE 5-6. PRODUCTION AND TRADE FLOWS, HYBRID CORN
SEED, LEADING EC COUNTRIES
(Metric Tons)**

1980/81	Production	Exports		Imports	
		Intra	Extra	Intra	Extra
France	95,629	19,526	12,912	63	3,098
Italy	19,521	89	432	1,294	8,498
Germany	2,565	700	251	18,507	11,078
1981/82					
France	119,017	20,239	13,408	665	3,193
Italy	18,822	636	131	1,130	12,651
Germany	4,967	581	404	18,338	12,705

Source: Eurostat.

tion to seed producers in other countries. Historically, the leading German seed firms, including Nordstaadt, Nungesser and KWS, have set up contracts with Eastern European farmers for their seed multiplication. This relationship is traditional, and the lower Danubian region, i.e., Hungary and Yugoslavia, have been German suppliers of seed for many years. The French made a big impact on the German market with the introduction of the early and very early varieties in the 1960s. INRA 258 was very important in helping to spread silage corn cultivation as far north as the Cologne region, and from this period, French breeders and producers became important suppliers of German corn seed. Until 1980–81, French exports of corn seed to the German market tended to increase, as revealed in the data on German imports shown in Table 5–7. The statistics show Germany's primary dependence on France as a source of supply of corn seed and secondary dependence on Hungary, Romania and Yugoslavia. The other suppliers, including Austria, Canada and the United States, were minor by comparison. In the late 1970s, the Eastern European producers lost market share as imports fell more than 2,000 metric tons. In the 1980s, the Eastern producers have re-established their position, with Romania ahead of Yugoslavia, and Hungary the overall leader. Throughout the period, France's share remained fairly steady, falling slightly from 52.7 percent in 1977–78 to 51.2 percent in 1979–83. As a share of imports, France reached a peak of 63.3 percent in 1979–80. This is a very heavy dependency on one producer, and the German market subsequently diversified from such concentration.

Beginning in 1977–78, the French have done well in the German corn seed market, but there may be serious problems ahead. Since

TABLE 5-7. CORN SEED IMPORTS INTO GERMANY
(Metric Tons)

	1977/78	1978/79	1979/80	1980/81	1981/82	1982/83 (est.)
France	14,110	16,330	15,648	17,751	17,781	17,250
Other EC	337	288	697	756	557	500
Total EC	14,447	16,618	16,345	18,507	18,338	17,750
Romania	944	1,380	1,843	2,251	2,693	3,250
Hungary	7,591	5,245	3,801	4,679	6,495	8,500
Yugoslavia	2,307	1,947	1,639	3,307	2,705	2,750
Total East	10,842	8,572	7,283	10,237	11,893	14,500
Austria	927	546	317	664	509	920
U.S.	521	406	107	53	313	440
Canada	—	—	653	124	—	90
Other	35	—	1	—	—	—
Total West	1,483	952	1,078	841	822	1,450
Total	26,772	26,142	24,706	29,585	31,053	33,700

Source: Eurostat.

1980–81, French export volume has declined even though total German imports have risen nearly 15 percent and, as cost figures in Table 5–4 showed, French corn seed production has a significantly higher cost structure than Eastern European seed producers. In 1982–83, this disadvantage ranged from 50 percent for single crosses to 100 percent for double and three-way crosses (attributable in large part to the CAP subsidy on grain corn). This is the reason for part of the problem and, given the perceived quality differences between French and Eastern European corn seed, the distortion caused by the CAP subsidy goes a long way toward explaining France's decline in price competitiveness. The logical response to these developments is for French companies to begin contracting out seed production to countries that seem to have a comparative advantage in seed multiplication. If the French agricultural system is so distorted by CAP subsidies that seed multiplication is uneconomic, then the appropriate strategy would seem to be movement of production to low cost areas. However, such a strategy creates problems for France and for most of the French companies. On the national level, it implies losing domestic production and exports, raising imports, and becoming more dependent on foreign suppliers for an essential agricultural input. At the firm level, French corn seed companies are reluctant to move corn seed multiplication abroad for their own reasons. Most are cooperatives where the farmers that own the operation as co-op members are also the farmers that do the seed multiplication. They have been benefiting from their seed plots for years and expect to continue receiving a multiple of the subsidized grain corn price for multiplying hybrid corn seed. They support the leader of the co-op, who is also usually the titular head of the seed operation. For years, this leader has succeeded in obtaining favorable agricultural policies and has protected the members from the effects of unfavorable developments in world markets. In such a situation, the migration of seed production is more than a commercial loss; it is a blow to the social, economic, political, and cultural fabric of local society. It is, therefore, not a step that can be taken easily or quickly.

A second reason for the reluctance of French seed companies to utilize more contract production of seed is their large sunk cost in production facilities in France. Because of transport expenses, it is costly to ship back to France the unfinished seed that has been multiplied in Eastern Europe. Thus, multiplication abroad might idle a significant part of the seed production plant and equipment owned by the companies. From the French perspective, this is a terrible dilemma that may in time weaken their whole corn seed industry. This serious concern has put the French industry at a crossroads, facing some unpleasant realities. In this situation, the initial reaction has been to try to protect the domestic corn seed industry by erecting trade barriers in the form of quotas on imports. However, before discussing these trade developments in detail, it is necessary to describe the evolving situation in the Italian corn seed market.

The Italian corn market is fundamentally different from the areas of the EC that lie beyond the Alps in that it uses late varieties of the type planted in the United States. As revealed in Table 5–5, the Italian corn crop is about 70 percent grain and only 30 percent silage. The silage corn is planted at higher elevations where there is not enough warmth for grain corn. The area Italy plants in grain corn is about 60 percent of the French area, and the Italian silage corn crop utilizes about 35 percent of the French area. Italy has about half of the total area of corn cultivation of France and about 50 percent more than Germany. During the late 1970s and early 1980s, Italy experienced a significant rise in imports and a consequent decline in domestic production of corn seed. This pattern is shown for Italy in Table 5–6 for 1980/81 and 1981/82, but the decline has been enduring. From its peak of 30,605 metric tons in 1978/79, Italian seed production fell to 12.2 m.t. in 1983/84. Imports rose from 8,210 m.t. to 13,160 m.t. in 1982/83, while estimated stocks of unsold seed rose from roughly 8,000 m.t. in 1978 to about 20,000 m.t. in 1984. There is no doubt that fundamental problems are accumulating that cannot easily or quickly be resolved.

The critical developments that have occurred in the Italian corn seed market, creating these problems, are:

- domestic firms have not introduced improved corn seed varieties in recent years;
- seed produced as public varieties in the United States can be imported cheaply into the Italian market;
- production costs in Italy are high due to the CAP subsidy program for grain corn;
- the domestic industry faces significant overcapacity problems and built up large stocks of unsold seed;
- there is a definite potential for increased imports of very low cost seed from Eastern Europe;
- the bulk of U.S. imports and much of the local production are a single variety.

Thus, the prospect is for declining income for domestic producers and insufficient resources to support breeding research programs needed to develop improved, genetically diverse varieties.

The varieties of corn seed available in Italy from domestic producers have not been noticeably improved for several years, and as a result, publicly available varieties developed by university breeding programs in the United States are competitive in terms of yield and other qualities. In fact, the most popular imported variety, known in the United States as B73 × MO17, is genetically identical to the preponderance of domestic Italian seed production. It is marketed under different names by different firms, and as a result in the mid-1980s roughly 60 percent of all Italian corn cultivation was dependent on a single variety. This is an undesirably high concentration level and implies potential genetic vulnerability of significant dimensions. Many small companies with agricultural marketing systems (e.g., pesticide and fertilizer distributors) have gone into the corn seed importing business. More than 50 small importers came into the market in the early 1980s and captured about 25 percent of the total market. These importers compete mainly by cutting prices and have no potential to conduct research or reduce genetic vulnerability. They are thus undermining the research capacity of the established Italian seed firms and are contributing to the increased genetic vulnerability of the Italian corn crop. They are supplying good quality, low priced seed to Italian farmers in the short run, but at a high long-term cost.

In addition to the absence of qualitative superiority, the Italian seed industry faces an acute problem in the form of high production costs. The single cross is the dominant hybrid in the Italian market, and estimates show that production costs were about 15 percent higher than French production costs in 1982–83. Referring to Table 5–4 for comparisons with importers, this implies that Italian production costs were 40 to 50 percent higher than U.S. costs and 80 to 100 percent higher than Eastern European costs in 1982–83. Since then, exchange rate movements in 1985 and 1986 have made things worse vis à vis the United States, and the potential for imports from the Eastern European countries remains.

From the Italian perspective, there is a clear and present danger to their domestic corn seed industry and the likelihood that the situation will worsen before it improves. Whereas France has one fundamental problem — high domestic seed production costs — Italy faces that and a second fundamental problem in the research capabilities of its domestic corn seed industry. It has not been able to generate improved varieties for its own market that differ in any way from varieties developed by U.S. public programs. And this failure has led to increased genetic vulnerability of the Italian corn crop through over-dependence on a single variety.

The initial inclination of Italian policymakers has been to restrict imports and protect the domestic corn seed industry. However, this policy tends to address the symptoms — declining domestic production and rising levels of imports — rather than the problems as just outlined. The imports are beneficial, at least in that they reduce prices paid by farmers and make it clear that basic problems exist. Restricting imports would raise prices without necessarily improving Italian genetic research programs. And protection against competition from imports would certainly not help to reduce production costs and make the industry more efficient. Nevertheless, increased protection has been the initial course of action.

RESTRICTING TRADE IN HYBRID CORN SEED

Beginning in 1982, the seed industries and governments in both France and Italy began to work for increased protection of corn seed production within the EC. The campaign for more protection against corn seed imports can be divided into three parts: the national phase, which occurred in the latter part of 1982; the EC phase, which took place in the first half of 1983; and the international phase, which happened in the second half of 1983. The driving forces behind the protectionist movement were the established French corn seed producers who faced stagnating or declining markets in the early 1980s. The largest Italian producers in the same predicament were their early and national allies in these efforts. By the time the data on the 1981/82 season were available, it was clear that after 1978, French exports of corn seed had stagnated for five years at about 33,300 metric tons. Growth in demand in the European market was going into increased imports from Eastern Europe in the case of Germany and from the United States in the case of Italy. Moreover, based on production costs, the evidence indicated a fundamental problem that would probably get worse with time. During the annual congress of the French Corn Growers Association at Aix-en-Provence in 1982, a motion on the need to limit imports was passed unanimously by the French corn seed producers. Within the French system, the private sector has a very large role in determining agricultural policy. In the area of corn seed policy, the key groups were the National Federation of Corn and Sorghum Producers (FNPSMS) and the broader National Interprofessional Group for Seeds and Plants (GNIS). The French corn producers soon met with several producers from Italy and Germany to explore the possibility of agreement on concerted action to limit imports. Despite a divergence in goals and priorities between the producers, there was general agreement to push for action by the EC under Article XIX, the Safeguard Clause, of the General Agreement on Tariffs and Trade.

The French producers had a singular goal of halting imports from Eastern Europe and making the EC countries the privileged customers of France. This view was well expressed in a 1982 article by the Secretary General of FNPSMS.[5] Within the EC is a regulation governing trade in corn seed (2358/71), the basic thrust of which is to establish a reference price for the importation of corn seed and to apply a compensatory tax to make up any difference between the reference price and the price of corn seed at the frontier. The reference price is to be the average of frontier prices over the last three years for certified, conditioned, treated seed in 25 kg. bags. The object is to keep out poor quality seed and to discourage price competition from importers. The French feel that the existing reference price has not kept up with the true rise in seed costs over the past 10 years and that adjustments to costs allowed for unfinished and unbagged seed have been abused. These, however, are peripheral issues compared with the 4 percent limit to the compensatory tax that is based on an agreement signed by the EC within the framework of GATT negotiations. The net result has been to render the EC regulation totally ineffective. To resolve this problem, the French and Italians recognized the need to enter GATT negotiations and have something to offer the other signatories, principally the United States, in return for reduced exports of corn seed. Under Article XIX of the GATT, compensation can be demanded by the parties that suffer as a result of the action limiting their exports.

The Italian producers wanted to forestall a growth in imports originating in Eastern Europe and to cut back on imports from the United States. Raising the 4 percent limitation would not stop U.S. imports because they are not greatly undercutting prices, but it would reduce them. It would help to stabilize prices and volumes in the domestic market and allow Italian firms to sell some stock. The primary targets of the Article XIX Safeguard action were to be Hungary and Romania, which did not affect the Italians' immediate problem, but Italy was willing to support France in the hope of obtaining some spillover effect on U.S. corn seed.

For the German companies, the situation was more complex. Several had long-term relationships with producers in Hungary and Romania, which they did not want to jeopardize. Other German firms were more closely tied to the French and were more supportive of protectionist action. On balance, however, the German seed companies and their association were opposed to attempts to restrict free trade in corn seed. The established German companies were concerned about another factor, the entry of new competition into their market based on low cost production of high quality seed. Pioneer Hi-Bred, which had brought superior varieties to the French market through France-Mais, has begun to establish an operation in Germany. It built a seed processing plant in eastern Austria near the Hungarian border and has production arrangements with seed multipliers in Hungary and Romania. This is not unlike the system used by German seed firms that contract out seed multiplication, but it is a potential threat to the status quo within the German market by the leading producer of high quality corn seed in the world. The current suppliers to the German market would probably like to freeze market share before Pioneer becomes a significant factor, but it is problematic whether the right regulatory package could be designed, or whether it would be accepted as the official German policy. For all these reasons, the German seed firms are more or less against a Safeguard action. The position of the German government is clearer — it wants a secure, diverse supply of appropriate high quality seed at reasonable prices. Since the 1978–81 period, the French share of the German market fell from 63 percent to about 50 percent in 1983–84. Eastern European seed rose from 29 percent to about 45 percent over the same period, not significantly different than the 40 percent supplied by Eastern Europe in 1977–78. From a German perspective, the increase in imports from Hungary and Romania is not necessarily a bad development. There is nothing sacrosanct about the French market share in the 1978–81 period and, as costs of multiplication rise in France, it is reasonable to expect some consequences.

By the end of 1982, the corn seed industry and the governments in France and Italy were committed to the pursuit of some protectionist action through the EC. This meant taking their case to the Common Market Secretariat in Brussels through the Common Market Seed Committee, the EC consultative group for the production and supply of high quality seed. Under pressure from the French and Italians, the EC tried to work out a position that would create as much common ground as possible and win approval from all the key members. One possibility was to renegotiate the tariff level, raising it to a higher level, but this was rejected. Renegotiating tariffs is a difficult and time-consuming process under the GATT, and it was not clear that a new tariff would provide the desired results. Furthermore, compensatory tariff reduction of a roughly equivalent amount would probably have to be made in another area. Another possibility was to invoke an anti-dumping action in which countervailing duties would be levied because the exporters were not pricing their product to represent its true costs. This would not help with U.S. corn seed and would present great problems in determining costs within the centrally planned economies of Eastern Europe. This approach was also rejected, and the decision was made to proceed with a Safeguard action based on Article XIX of the GATT.

The GATT was organized in 1947 as a tariff reduction agreement among 23 nations. Over the years it has evolved to become the forum in which leading countries negotiate on trade matters, establish rules and procedures governing trade, and oversee the performance of the trading system. The objectives of the GATT are to promote an orderly and expanding flow of traded goods (and services) worldwide. It seeks to do this through the elimination of quotas and other quantitative restrictions on trade, the reduction of tariffs, and the application of the principle of nondiscrimination. Since 1947, seven rounds of tariff reductions have been negotiated under the auspices of the GATT, and the general procedures governing world trade have been set out in articles that the signatory countries have agreed to support. Under the

GATT, countries facing economic problems resulting from international trade have recourse to several types of exemptions or escape clauses. Article XIX is one of the most flexible in terms of its potential to offer relief to countries facing import-based problems and, as with most legalistic systems, precedent is an important factor in justifying an action under this article.[6] Because this article has the potential to provide broad relief on imports for GATT members, actions taken under its auspices are called Safeguard actions.

The five criteria that must be met for a trade problem to qualify for relief under Article XIX are that (1) unforeseen developments, (2) resulting from one or more GATT obligations, (3) due to imports in increasing quantities, (4) threaten or cause serious injury to domestic producers (5) of like or directly competitive products. This also is generally considered to extend to GATT members that have a preference in the importing country where loss of market threatens serious injury to domestic producers in terms of lost exports. The basic remedy is to withdraw or modify the GATT obligation supposedly causing the import problem. Invocation of Article XIX first requires written notice of intent, with the opportunity for the affected parties to consult except where irreparable damage would be caused by delay. The concession on the GATT obligation is supposed to be temporary, lasting only as long as necessary to resolve the problem. And all actions are to be done on a nondiscriminatory basis. The last point is currently a matter of serious disagreement between some EC countries and other GATT members. Therefore, applying Article XIX in a discriminatory way—against one or a few countries—is a very grave step, and it is understood that compensatory remedies are available to all the affected countries in terms of lower tariffs from the country invoking the Safeguard action, or higher tariffs aimed against exports from the invoking country. There is the presumption throughout that the unforeseen disruption and the resulting imports are not permanent and that the threatened industry can recover and become competitive again. Where the change in competitive position is permanent, Article XIX would seem to imply a phased withdrawal of the invoked action. Article XIX is not intended to provide permanent protection in a situation where comparative advantages have shifted, but room to recover from temporary and unforeseen developments.

Applying these general principles involving Article XIX to the corn seed problem facing the EC raises two interesting and important issues. Beginning with the criteria, whether something could or could not have been foreseen can be argued endlessly, and the case will not rise or fall on this matter. The rest of the criteria also probably apply, except for causation. Specifically, the 4 percent tariff has been in place for many years and was not an encumbering factor in the past. Until about 1978, French production and exports grew steadily despite the presence of the 4 percent tariff on imports of corn into the EC. Therefore, the causation part of the criteria is weak at best. Clearly, something else has changed that is causing the problem, and that is the rising costs involved in the production of seed in the EC. And in Italy, as noted, there is an additional factor, the failure of domestic breeders to develop new improved varieties over the past decade. The threatened or actual damage cannot be attributed solely to the 4 percent tariff, i.e., the GATT obligation; a significant share of the problem is attributable to EC policies, i.e., the CAP subsidies on grain corn, and the failure of the Italian firms and perhaps some of the French firms to develop new superior varieties. The varieties available from Eastern Europe and public sources in the United States are qualitatively competitive and cheaper to buy, whereas 10 years ago, French and Italian firms were able to beat the competition on quality without too great a cost differential. The determinant changes have occurred within the respective countries, not in the GATT obligation. Therefore, the degree to which the problem resulted from the tariff have to be established.

The second issue that would have to be clarified concerns the permanence of the Safeguard action. Based on available evidence, the competitive position of the EC seed producers is not going to improve until the distortional effect of the CAP subsidy on grain corn is removed. There is no current evidence that this will occur, so at the very least, a schedule for the phased withdrawal of the Safeguard action should be included in the invocation itself. Where much of the problem is due to actions of the government, in this case the CAP, and failings of the industry to adequately research and continually improve the product, it is not clear that exemption from GATT rules is justified. More important, from a policy point of view, it is unlikely that an action under Article XIX will solve the problem unless accompanied by reform of the CAP and more extensive and more serious genetic research by the leading European corn seed producers. In the absence of these reforms, relief from the 4 percent tariff limit could cause European farmers to pay high prices for the same quality seed that can be bought for less elsewhere.

Although work on an action based on Article XIX reached advanced state within the Common Market Secretariat in the late spring of 1983, this idea was finally discarded. A Safeguard action was rejected for several reasons. The EC cannot move on an issue unless there is agreement among the key countries, and Germany was not in favor of formal action at that time. And because Article XIX is supposed to be nondiscriminatory, action could not easily be directed against Eastern Europe without also affecting U.S. exports to Italy. In addition, given the difficult state of American-EC relations in the agricultural area, the Europeans were not willing to provoke the Americans over this issue. Therefore, a formal Safeguard action—which the French seed producers clearly wanted—was rejected in favor of the much milder step of requesting negotiations on a trade problem.

The overall state of trade relations and the importance of the respective countries to each other are significant considerations in this issue. Hungary, Romania and the Eastern bloc countries in general are of less importance to France than to Germany. These countries purchase more exports and have borrowed more from Germany than from France. The Germans have been much more active in promoting good political relations with their Eastern neighbors than the more westerly members of the EC and are, therefore, more reluctant to approve any action that would aggravate commercial relations with the countries of Eastern Europe. Given that Germany's interests were not advanced by a Safeguard action and that the case for such action was not iron-clad, it is not surprising that the Germans preferred to defer formal invocation of Article XIX.

Even more important is the state of relations between the EC and the United States over agricultural policies. Because of various effects of the CAP and an increased aggressiveness in U.S. international agricultural policy, the relations between the EC and the United States are passing through a particularly difficult period.[7] It is a broad conflict that impacts on interests much more important and powerful than corn seed. With high level efforts to improve these relations under way, the EC Secretariat was not anxious to provoke the Americans by invoking a Safeguard action over imports of corn seed. There was some talk of getting the United States to accept corn seed quotas in return for a more generous quota in corn gluten feed, but the United States rejected both the carrot and the stick. The failure of this tradeoff forced the EC to confront the prospect of further irritating the United States, and the EC backed off. The issues were not sufficiently clearcut, the interests sufficiently large, and the EC sufficiently unified to justify a confrontation with the United States over the EC's corn seed problem.

The other countries—the United Kingdom, the Netherlands, Belgium, Denmark, and Greece—were not in the center of the conflict and ultimately agreed to the solution arrived at by the three large countries that have 90 percent of the corn-growing area in the EC. Among the remaining countries, Greece probably sided with France; Belgium and the Netherlands probably saw their interests as similar to Germany's; and the United Kingdom preferred a move toward greater reliance on price solutions, i.e., raising the levy now frozen at a 4 percent ceiling by a long standing GATT agreement. Negotiations on the levy would be difficult and drawn out, however, so none of the others were anxious to pursue this course.

After dropping the proposed Article XIX action, the EC held meetings with each country involved—Hungary, Romania and the United States—to determine the nature of the problem existing between the

EC and each of the exporters and to work out some sort of bilateral arrangement to limit exports. Such arrangements are called Voluntary Export Restraints (VERs) and have become fairly common in recent years.[8] These meetings did not advance very far, and while some increased awareness of the EC's problems may have occurred, there were no agreements by any of the parties to restrict exports. In early 1984, the issue began to recede as an international policy matter, but within the EC there is continuing concern.

FOOTNOTES, CHAPTER 5

1. I am indebted to Edgar Rihm of Semillas Pioneer, S.A., Seville, Spain for the historical data on corn yields.
2. However, some feel that with LG–11, Limagrain should have been able to capture an even larger share of the German market than it did.
3. *French Seeds* (Paris: GNIS, 1982), p. 9.
4. Maturity numbers differ between Germany and France by about 30 points. The French system, which has lower numbers for the same hybrid, is used here.
5. Michel Tiger, "The Exchanges: An Ineffective Community Regulation," *AgroMais* (December 1982), No. 13, pp. 46–47.
6. For more on Article XIX and other aspects of the GATT from the legal perspective, see John H. Jackson, *World Trade and the Law of GATT* (Indianapolis: Bobbs-Merrill, 1969).
7. For a discussion of this see Stefan Tangermann, "What is Different About European Agricultural Protectionisms," *The World Economy* (March 1983), Vol. 6, No. 1, p. 39.
8. For more on VERs, see Brian Hindley, "Voluntary Export Restraints and the GATT's Main Escape Clause," *The World Economy* (November 1980), Vol. 3, No. 3, p. 313.

Seeds and the Developing Countries

6

THE AGRICULTURAL SETTING

Chapter 1, "Seeds and the World Food Balance," described the overall agricultural situation of the developing countries. In the 1950s and 1960s, economic development usually implied a focus of resources on industrialization, and agriculture was not adequately emphasized. Food prices were kept low to hold down costs for urban workers and to encourage industrial expansion. To support this strategy, imports of foodstuffs rose dramatically; these were often on concessional terms and acted to depress local food prices.

On the input side, agriculture was also discouraged. Investment levels were low and the costs of inputs, particularly fertilizer and pesticides, were very high in terms of prices received by farmers for agricultural produce. The result was slow growth in agricultural output and increasing dependency on food imports. By the early 1960s, it was becoming evident that this policy could only lead to a progressively deteriorating situation. Surplus world stockpiles were declining, and concessionary food could not be counted on as a permanent solution. Within the domestic agricultural sectors was ample evidence of the failure of policies that shortchanged agriculture: rural poverty, landless peasants, inadequate production of agricultural raw materials for the industrial sector, and a failure of food production to keep up with expanding domestic demand.

Agricultural policies changed slowly, and various regions reacted in different ways. Africa expanded cereal production rather well in the 1960s, but this performance fell off sharply in the 1970s. Cereal output grew less than half as fast as population and, even though per capita food consumption fell, imports nevertheless rose at an annual rate of 10 percent in the 1970s. It is in Asia that the greatest improvement was achieved. Problems in the 1960s were especially acute on the Indian subcontinent, but after the mid-1960s, agricultural production began to accelerate. There was greater emphasis on investment for irrigation, more use of fertilizer and the introduction of high yielding wheat and rice varieties. In Asia as a whole, the situation improved notably in the mid- and late 1970s. As a result, imports into Asia leveled off and began to decline in the 1960s and then fell significantly in the 1970s.

As a group the middle income countries, including both oil exporters and oil importers, experienced rapid and consistent growth in cereal output over the period 1960–80. However, the combination of population growth, higher incomes and a strong desire to improve individual diets pushed demand up even faster. Most of these countries had foreign exchange available and, as a consequence, food imports rose at an annual rate of 6 percent in the 1960s and 10 percent in the 1970s. These imports were financed by expanding exports of petroleum (oil exporters), of manufactured goods (newly industrializing countries), and international lending. This pattern continued through 1980, but since then reduced demand for exports, a falloff in international lending and the cumulating balance of payments difficulties have forced a cutback in food imports among many middle income countries. In the 1980s, the need for more rapid growth in domestic agricultural output is critical, particularly among the middle income developing countries facing external debt problems and the African countries facing severe food deficits.

Despite the successes of the 1960s and 1970s, there remains enormous potential for further gains in agricultural output and improvements in agricultural policies. Although public sector spending has led to increases in irrigated land, rural infrastructure and the beginnings of a research capacity, it is still very low—totaling only 5 to 10 percent of most countries' budgets.[1] Moreover, much of the money spent sustains parastatal agricultural organizations (to be discussed later) that are inefficient at delivering the services and products they are supposed to provide for the farmer. In many cases, these parastatal organizations are legal monopolies and competition is prohibited. Even when private activity is allowed, the parastatal usually enjoys key advantages such as preferred access to imports, subsidies in its operations and special relationships with other government agencies. In some cases, the parastatal and regulatory agencies have been so inefficient and obstructionist that their efforts have been counterproductive to agricultural development. Governments are beginning to recognize these problems, and some have taken steps to improve the efficiency of the agricultural system by freeing markets and allowing private companies to compete. To date such steps have been halting and infrequent, but they have begun to have a significant effect within a select group of countries.

SEEDS IN THE CONTEXT OF AGRICULTURAL DEVELOPMENT

The list of problems facing agriculture in developing countries is long and challenging and includes credit, education, land tenure, transportation, institutional and cultural obstacles, marketing, weather, energy costs, and many others. Only a few will be mentioned because they relate directly to agricultural seeds and seed policy. It will be apparent after considering these problems that seeds will be more important than ever in achieving satisfactory progress in Third World agriculture.

Expansion of cultivated land will be a declining factor in agricultural development in the years ahead. Between 1960 and 1980, expansion of land accounted for almost 20 percent of agricultural growth in the developing countries, but is projected to account for only about 10 percent of their agricultural growth between 1980 and 2000.[2] Theoretically, a great deal of arable land remains available for cultivation, but almost all of it is unsuitable because of remoteness, disease or potential deterioration in the ecological systems of the regions. Most countries face severe limits on the availability of suitable new agricultural land, and even those with frontier regions, e.g., Brazil, usually face enormous infrastructure costs before the new lands can be incorporated into the market economy of the country. Without roads, communications and public facilities, the newly settled lands cannot be used for much besides absorbing surplus population in subsistence agriculture. Therefore, land expansion will remain a small part of overall agricultural productivity; the preponderant part of the solution must come from improved yields.

Crop yields in the developing world have been rising by 2.2 percent annually since 1960.[3] In areas where appropriate development policies have been applied, including irrigation, fertilizer and high yielding seed, yields have doubled in a decade, with an annual compound growth rate of 7 percent.[4] Irrigation has been particularly important in increasing agricultural yields in Asia; it can permit double cropping, and practically eliminate the risk of crop failure due to drought. The problem with irrigation is that it is expensive to put in place, commonly requiring fixed costs of $2,000 to $3,000 per hectare (and additional variable costs) for a well designed project, and it can only be applied where surface or well water provides adequate supplies (note that the water is a variable cost). Irrigation will continue to be utilized where natural resources and expected returns make it feasible to invest the capital required. Private irrigation schemes utilizing tube wells and small pumps have become common in some sectors of India. These are more expensive to operate and maintain per unit of land irrigated, but the initial investment is lower, the charge to public funds is much reduced, and the individual farmers manage and maintain the system. This is an option that deserves consideration when it presents a feasible alternative to large-scale publicly funded irrigation projects.

In examining irrigation in the developing world, two important realities emerge: in many instances, the most attractive irrigation investments are already being exploited, and the overwhelming majority of farmers in Asia, Africa and Latin America live in rain-fed agricultural areas. There are some areas where viable new irrigation programs can be established, but the emphasis in agricultural development must shift to rain-fed regions because the bulk of Third World farmers are located there. These areas are not generally suitable for irrigation, and yield gains will have to be based on cultivation practices appropriate to rain-fed conditions. Moreover, local conditions tend to vary enormously, so that correct cultivation practices — in terms of tilling, crop rotation, soil conservation, drainage, and so forth — are specific to a given area and cannot always be successfully deployed elsewhere. In irrigation-fed areas, the problems of agricultural development are amenable to capital-intensive solutions, but rain-fed areas require management and knowledge-intensive strategies — institutional change, legal reform, improved extension services, efficient markets, and appropriate policies. These issues are moving to the forefront of agricultural development policy.

The challenge of development in rain-fed agricultural areas is different and likely to be more difficult than the problems overcome by successful irrigation programs of the recent past. Because capital is less important, leverage of the international development institutions may be reduced. The "carrot" of finance for large-scale high visibility projects is not as strong, and the countries are thus likely to be less willing to undertake difficult reforms and policy adjustments. At the same time, because more farmers live in rain-fed areas, the necessary changes will affect more people than was the case with irrigation-oriented policies. This, in turn, suggests that the short-term political and social costs will be higher for agricultural development strategies aimed at rain-fed regions.

Special Conditions Facing Small Farmers

Another factor compounding these difficulties is the prevalence of small farmers in the rain-fed regions. The basis of past successes has frequently been the introduction of new agricultural practices among more advanced farmers in the leading agricultural regions. Other farmers then tended to emulate the successful practices of their neighbors. However, the overall impact of these programs has been strongest among the relatively larger farmers holding land in regions suitable for irrigation. By comparison, the smaller farmers in the rain-fed regions are less advanced to start with, have less capital to invest and are not as well educated. Consequently, the focus of agricultural development must shift and concentrate more on the practices and problems of the small farmers.

For the 1980s and 1990s, improved seed is perhaps the most important input through which agricultural yields and incomes can be raised for the small farmers in rain-fed regions. However, because of the conditions facing them, small farmers are reluctant to change their established patterns of cultivation.

One important reason for the slowness of small farmers to utilize improved varieties has been the absence of a dramatic genetic breakthrough comparable to the dwarf rice and wheat varieties, which helped raise yields in irrigated regions. The improved seed varieties suitable for rain-fed areas have not been as predictably superior to the old varieties. The yield gains available to rain-fed farmers are very attractive under favorable moisture conditions, but the gains are frequently smaller, percentage-wise, under poor moisture conditions. For the small farmer, such variability is unacceptable, especially for the subsistence crop that the family needs to survive.

Another factor is the wide range in local conditions facing farmers in the rain-fed regions of Asia, Africa and Latin America. Where water is available, adjustments for variations in local conditions can be made through the application of other inputs, especially fertilizer. In drier regions, fertilizer applications may not be as successful, because fertilizer and water are complementary. Without adequate water supplies, fertilizer is not as effective. Thus, taking the risk of low rainfall into account, the superiority of the new variety of seed over the traditional variety has usually not been enough to convince small farmers. Varieties that provide high and stable harvests within a wide range of moisture conditions would constitute for them a dramatic improvement comparable to the rice and wheat seed improvements for farmers in irrigated areas in the 1960s and 1970s.

Over time, the proportionate aggregate yield gains likely in rain-fed agriculture can be expected to be comparable to the experience in irrigated areas. But because progress is likely to be more irregular, there is an important potential conflict. From the farmer's point of view, the risks inherent in adapting new seed are seen to be high, while the immediate payoff seems modest. From the point of view of the policymaker, the medium- to long-term payoff from improved seed is unmistakable. Moreover, the risks are manageable because they are spread over time and across regions. One objective for policy ought to be risk reduction for small farmers; if individual risk is perceived as being higher than social risk, then some method for reducing the risk facing individuals should be found. The result would be a congruence of individual and social risk perceptions, more willingness on the part of small farmers to accept new approaches, and a high aggregate level of output.

The low absolute level of output that currently prevails in rain-fed agriculture is another consideration from the policymaker's perspective. A low starting point makes large proportionate gains possible, even though absolute output per unit of land will never approach that of irrigated areas. The stakes in successfully introducing improved seed into rain-fed agriculture are nevertheless very high, probably greater than for the high yielding varieties in the 1960s. This is because of the vast amount of land in rain-fed regions and the large numbers of extremely poor people who would be affected. Through a continuing series of improvements in seed and some adaptation to new agricultural practices, significant gains in output can be achieved in rain-fed regions. While results would vary in different locations, experts feel that an increase in yields of 50 percent between the mid-1980s and mid-1990s is feasible and that gains of 100 percent for certain regions are entirely possible. Such gains in output not only will help the world food balance, but also will improve the diets and increase the incomes of many of the poorest people in the world. Increased output in rain-fed agriculture is the most welcome kind of economic growth in that it simultaneously raises the world's income and improves its distribution.

While it is true that the genetic breakthroughs have not been as dramatic for farmers in rain-fed regions of developing countries as for other farmers, it is also fair to say that a large part of the problem of getting improved seed accepted is on the demand side. Even when clearly superior varieties have been available, small farmers have been slow to use improved seed. If the seed does not sell itself as effectively as the high yielding varieties, the seed suppliers in the rain-fed regions will have to be better at distribution, promotion and extension services. More effort must go into understanding the problems of the small farmer and working on seed varieties and agricultural practices perceived by small farmers to be solutions to their problems. Small farmer reluctance to adopt new agricultural inputs and techniques is perhaps the most critical challenge facing agricultural policymakers today.

Four key determinants of small farmer behavior can be noted:
(1) the purpose for which the crop is being grown;
(2) the state of development of local infrastructure, particularly the transportation network and the marketing system;
(3) the land tenure system;
(4) the farmer's education level.

Experience has shown that farmers are highly risk-averse when the crop at stake is the staple of their diet. However, when the new seed is for a cash crop and the farmers' subsistence is not directly involved, there is much more willingness to experiment with new products or new techniques. Transportation is particularly important because it is frequently the determinant of whether markets are accessible. The markup or margin charge on both inputs and output is much less when transportation is well developed. Good transportation in turn leads to much greater potential for marketing a cash crop, supplying low cost inputs and developing the agricultural system. Land tenure arrangements are of course critical. Development becomes very difficult if holdings are prohibitively small, or if tenants do not have adequate incentives to save, to take some risks and to increase output. When all these factors are negative, i.e., tenant farmers growing subsistence crops in an inaccessible area, it becomes increasingly difficult to make any headway. Finally, education is almost universally low among small farmers because of the typically difficult circumstances into which they are born. However, when some farmers can be found with even a little training, it may be possible to work with them and reach others through their example. Generally, though, education and capital (i.e., financial savings, equipment, animals, structures, and land) are in very short supply among small farmers, and development policies must take these shortages into account. From almost every aspect, therefore, the challenges facing agricultural development are greater in the rain-fed regions than in the irrigated areas.

Seed Development Strategies for Small Farmers

To begin with, the small farmers' basis of evaluation of new varieties tends to differ from that of plant breeders or seed producers. In general, small farmers do not fully understand the reasons for their success or failure. Those that have survived have established patterns and procedures that have worked in their situation. Tradition is a time-proven means of reducing risk to a manageable level and increasing the chances for survival. Saving one's own seed and selecting for stability of yield are examples of functional traditions that have helped small farmers endure for centuries. For many staple crops, e.g., corn, the amount of seed that needs to be saved is small and easily stored. Another functional tradition is to be as independent as possible, keeping control over many of the critical inputs. Such traditional practices are unfortunately now an obstacle to going beyond survival to the achievement of a small measure of prosperity in the short term and significant gains in the medium term.

It is worth noting five critical areas of differing perceptions between small farmers and many breeders regarding desirable genetic characteristics in staple crops:[5]

(1) yield: the small farmer seeks a good but stable yield, while the geneticist working under better conditions has, in the past, emphasized maximum yield;

(2) grain quality: small farmers are usually quite interested in specific factors (which may differ among small farmers) such as taste, appearance and storage characteristics under local conditions, whereas breeders still do not give enough weight to these factors;

(3) plant structure: the small farmer may be interested in the stalks and foliage of the plant for animal feed, in addition to the grain yield, which is the primary concern of breeders; (4) cropping practices: the small farmer often mixes or alternates crops and uses manual techniques, while the breeder tends to think in terms of mechanization and a single crop; (5) seed appearance: small farmers tend to select their own seed based on good size and appearance.

There is no substitute for knowing what farmers want and what will convince them to try a new variety of seed. Links between farmers and plant breeders must be created by the extension system and through seed producer demonstration plots. Breeders must know what farmers want and farmers must know what breeders are making available. This kind of farmer-oriented program is somewhat novel in the developing world, and experience is only now being accumulated as to how to make it work.

In seeking to provide improved seed to small farmers, a reasonable strategy ought to take into account the following considerations. Seed programs for small farmers should be launched in areas where success is most likely and then should shift to more difficult regions as experience and confidence build up. This means that the seed program should focus first on areas with an existing transportation network to enable increased output to be moved to market at a reasonable cost. If it is possible, the program should begin with nonsubsistence crops. Seed production should emphasize stability and security as much as yields; this includes stability of yields under various conditions, assured availability of supplies for the farmer, and the maintenance of seed quality as production of improved seed rises. Development of a good marketing system must have a high priority, and an incentive system must be established to encourage distributors to promote improved seed. One critical aspect of marketing to small farmers is to make the seed available in smaller package sizes appropriate for typical small holdings. Moreover, the package should be sealed so that deterioration during storage and transportation is minimal and the farmer is assured the bag contains the improved seed he intended to buy. Seed should be introduced as a part of an integrated package of inputs, including fertilizer and new agricultural techniques. All the agencies — the extension service, the farmers' associations, the firms supplying inputs — should be coordinated in this effort to communicate the same message to the farmer.

These considerations apply very widely, if not universally, but emphasis and priorities will vary depending on national and local conditions. Seed development strategies for small farmers must be carefully designed with the strengths and weaknesses of the country's agricultural system and seed industry clearly in mind. Progress will be slow and must be properly founded if it is to be sustained. Attempts at quick fixes are likely to make longer-term problems more intractable in the future.

SEED INDUSTRY DEVELOPMENT

As was argued in Chapter 1, genetic research and improved seed have become more important than ever to agricultural progress. In developing countries, there are still some gains to be made from expanding the area of production, but usually at a high cost in infrastructure. Energy-based inputs will continue to expand, but at a slower rate due to their cost. Institutional and cultural change is essential, but is usually a slow and difficult process. Seeds are, therefore, an increasingly important part of agricultural development strategies, and governments are coming to recognize that their seed industries must be strengthened if they are to meet the challenges before them. A list of the objectives of plant breeders and the seed industry in developing countries clearly demonstrates these challenges:

(1) greater water-use efficiency, with particular application to rain-fed areas;

(2) increased yields, while adding to reliability and security;

(3) improved energy efficiency, implying pest resistance, nitrate assimilation and low use of chemical inputs;

(4) higher harvesting index;

(5) enhanced nutritional value;

(6) ability to be integrated into a program that can be managed by small farmers;

(7) diversity in the current varieties utilized to reduce genetic risks;

(8) improved replacement varieties coming on-stream every three to five years;

(9) timely provision of appropriate seed supplies, even in remote areas;

(10) a modest overall cost.

This is indeed a tall order, but most of the above goals will have to be met if agricultural progress in the developing world is to continue at satisfactory rates through the 1980s and 1990s.

Almost universally throughout the developing world, the seed industry is currently inadequate to achieve the goals just outlined. In the 1960s and early 1970s, agriculture in the developing countries had two great advantages that are no longer available: cheap energy based on low cost Middle Eastern petroleum, and dramatic genetic advances in two key crops coming from the International Agricultural Research

Centers (IARC). Low cost energy is most unlikely to be available any time soon. Genetic research will continue to make progress in developing new improved varieties, but a major breakthrough—as occurred as a result of the dwarfism genes introduced in wheat and rice varieties —is unlikely to be available to world agriculture before the late 1990s. By then, bioengineering products could begin to make a notable impact, but until that time genetic progress probably will be incremental. With the two advantages just noted, progress in agriculture in the developing world in the 1960s and early 1970s occurred despite the presence of generally inappropriate policies. The performance was worst in Africa, which had particularly poor policies, extended drought in large areas and less land where the high yielding seed could be applied.

In the 1960s in most developing countries, policies tended to depress the price of agricultural production of foodstuffs to hold down costs to consumers in urban areas. As described above, this had the unwelcome side effect of discouraging local agricultural production and forcing the country to import. One of the countermeasures taken to offset low prices for agricultural output was to hold down the prices of some agricultural inputs. Seed was usually a prime candidate because, unlike agricultural chemicals, it could almost always be produced domestically. To implement this policy, one or more parastatal organizations were set up to help provide the country's seed requirements at a low price to the farmer. Over time, these parastatal seed organizations often evolved into sizable bureaucracies receiving the bulk of their financial support through the government budgetary system rather than through the provision of services to farmers and agricultural producers. In some countries, these organizations were established as legal monopolies, but in nearly all cases they had an advantageous position vis à vis the private sector. The first advantage was typically that part or all of their costs were subsidized by public revenues. A second common advantage was a special relationship between public sector seed production activities and the seed regulation and certification agencies. A third advantage was their relationship with the extension service providing advice and technical support to farmers. A fourth advantage was preferential access to new varieties developed by the International Agricultural Research Centers. In some cases, the public sector seed organizations also had preferred access to foreign exchange and priority in the importation of equipment and supplies, and these organizations were usually exempt from various domestic taxes. Even when the public sector did not have a legal monopoly, the result was to weaken the private sector to the extent that meaningful competition could not develop. Under these conditions, the parastatal seed operations frequently became high cost inefficient suppliers.

New entrants to the market can be effectively excluded by the combination of advantages enjoyed by the public sector. The subsidization of costs enables the parastatal seed organizations to sell at very low prices despite their inefficiency. This acts as a significant barrier to new entrants who must cover their costs (including some taxes) through sales of their products. No matter how efficient they may be, private seed firms find it very difficult to enter such a market and compete effectively. The start-up costs of establishing a seed operation tend to be high, with a long pay-back period, under the best of circumstances. When the main competition is heavily subsidized and selling below cost, it is extremely hard for a private investor to justify the expense and risk required. Compounding the economic problems facing private seed companies is that their operations are usually regulated and their varieties certified by agencies closely related to, or sometimes part of, the parastatal seed producer. At the distribution end, a similar problem exists because seed is frequently supplied to farmers through public sector agencies that are part of, or closely aligned with, the public sector seed producing organization. The private company therefore has to create a distribution and support network or hope that the existing system will effectively promote and distribute the firm's products. Finally, private companies face increasing limitations on their access to imported equipment, parent (breeding) seed and materials. Taken together, these disadvantages have so severely weakened the private sector that, probably in the majority of the developing countries, there is little or no effective competition among seed producers and consequently no seed industry in any meaningful sense of the term.

Government policies in the developing world have typically created a situation in which an inefficient public sector seed organization dominates, the local private companies are small struggling entities, and the international seed companies that are present have suboptimal operations that cannot properly contribute to the agricultural development of the country. The net result is a national system of seed supply that is inadequate to meet the expanding agricultural requirements of the country.

There is no question that in the early stages of seed industry development the public sector has to take the lead in the provision of seed to farmers. The public sector has widely demonstrated its capacity to perform research, certification, quality control, regulatory, and extension functions. These are areas of responsibility in which the public sector necessarily has a critical role to play. At the operational level, the public sector seed organization typically has an innate disadvantage because it has little or no economic inducement to serve the farmer. Although most of the staff in these organizations are well motivated, especially the breeders conducting the research, extensive teamwork is essential to make a success of seed operations. The failure or tardiness of only a few people—in delivering the seed, for example—can undermine the efforts of all the others. There are so many critical operations in producing seed that failure in any number of ways can render the overall effort worthless. If this happens in a private company, the effect is felt immediately and unmistakably throughout the organization. But in a parastatal organization where the preponderance of revenues comes from the government, it is very rare for people to lose their jobs or have their pay reduced because farmers did not receive the right seed at the right place at the right time.

During the 1960s and 1970s when the high yielding varieties were being introduced, the new seed was clearly superior to the varieties then in use. Once this was recognized by the farmer, demand for the seed overwhelmed the system, and problems on the supply side were not a prohibitive deterrent to progress. The seed was such an advance that the agencies and organizations involved in its production and distribution tended to enjoy success through association. Other factors contributed to the appearance of success and competence. The high yielding varieties were best suited to specific regions, and they met their initial acceptance among the more advanced farmers, which simplified the seed producers' job in terms of physically distributing the seed and follow-up support. In the 1980s and 1990s, as noted above, the genetic advances are likely to be in smaller absolute increments, so the new seed may not be as superior relative to the currently utilized varieties. In proportional terms, the gain may be as great because of very low current yields. But the farmers that must be reached are less advanced, will need more support and are spread out over vaster areas; therefore, the probability of realizing the full potential of the new seed is lower. Finally, the risk is greater in the rain-fed regions where development efforts must now be focused; good seed could get a bad reputation, for example, if rainfall is inadequate in the year improved seed is introduced and the native varieties perform nearly as well. The net result of these factors is that seed producers in developing countries today face a more difficult challenge than in the 1960s and early 1970s; most notably, performance will have to improve in the distribution and "customer service" areas of the operation if seed is to play its designated role in agricultural development.

A Four-stage Model of Seed Industry Development

It would be useful here to outline a model of seed industry development to help classify and analyze the progress of the seed industry in different countries. This is a course of evolution that countries seem to follow, although there are certainly variations of the central pattern. The four stages of agricultural and seed development as identified by Johnson Douglas[6] are as follows:

(1) Agriculture is predominantly subsistence, seed research and development programs are not effective, and seed is almost entirely farmer-saved.

(2) Seed research and development is under way, improved seed has begun to replace traditional varieties, some commercial seed multiplication is taking place, fertilizer and other inputs are increasingly used, and the lack of sufficient improved seed is a constraint to development.

(3) Seed research and development is established, high yielding varieties are replacing traditional seed, fertilizer and other inputs are widely used, seed quantity is generally adequate but distribution is inefficient, and only a few private firms have emerged.

(4) Advanced agriculture predominates in most regions, private commercial seed production and marketing are becoming stronger, effective seed laws are in place, and an integrated national seed industry with international links is developing.

Douglas notes that within a country these stages will vary for different crops and regions.

It is possible to fit countries into levels of seed progress based on these four stages. In Asia and Latin America, almost all countries are beyond Stage 1 and in either Stages 2 or 3, depending upon the crop and the country. In Africa, many countries are still predominantly in Stage 1 with a few in Stages 2 or 3, and then not for all crops. There are several developing countries that have reached Stage 4 for one or more crops, but this might be disputed. Agricultural and seed production have generally advanced to Stage 3 in regions where high yielding varieties have made an impact. This has usually occurred in irrigated land or in areas with very reliable rainfall. The greater part of rain-fed agricultural lands are in Stages 1 or 2 and must be brought up to at least Stage 3 if progress in agricultural development is to be sustained. In addition, many of the crops and regions now in Stage 3 of seed development must progress to Stage 4 during the next decade to maintain satisfactory progress in agricultural output for the country as a whole.

In most instances where seed production in developing countries has progressed to Stage 3, it has been based on high yielding varieties of wheat and rice, new export crops such as soybeans, or on the availability of suitable hybrid corn seed. This has been achieved despite weak performance by the seed producers in many operations. Progress was made in large part because of the desire of the farmers to attain superior seed. In the absence of genetic breakthroughs applicable to rain-fed agriculture, an alternative to Stage 3 may evolve. To bring rain-fed agriculture into Stage 3, it may be necessary to improve the domestic seed industry beyond the level that was needed to bring irrigated agriculture to the same stage. If dramatically improved varieties are not available, either domestically or through imports, to make up for deficiencies elsewhere in the seed production system, then a country may remain in Stage 2 until the deficiencies in the local seed industry are corrected. As a result of the basic research of the IARCs, the performance on the research end of developing countries' seed operations has been adequate to good. However, as the product has moved toward the ultimate user, performance has fallen off significantly. The seed industry's operations and activities that have presented the greatest problems in the developing countries are:

- certification or approval of new varieties;
- multiplying out new varieties;
- adequate technical training;
- maintaining quality control in seed production;
- timely distribution of appropriate seed;
- financing that encourages optimal agronomic practices;
- advising farmers on obtaining the best results from new seed varieties;
- learning what new varieties and products farmers need.

Developing countries must make progress in correcting deficiencies in these operations over the next few years; failure to do so will seriously threaten future gains in their agricultural sectors.

These are primarily operations that have been carried out by private seed companies in the developed countries, most of which have been in Stage 4 for a decade or more. To date, most developing countries have manifested a reticence to implement the kinds of reform that would permit private seed companies to grow and make a significant contribution to local agricultural development. At the same time, the parastatal organizations set up to provide improved seed have not yet found a successful formula, particularly for quality control, distribution and follow-up activities. If such a formula cannot be found, and if seed research for rain-fed agriculture does progress in small incremental steps rather than a dramatic breakthrough, then developing countries may find their existing seed policies have led to a dead end. This line of argument would suggest that a new approach to the issue of seed production may be in order—an approach that will strengthen private seed companies while encouraging them to take over the seed operations and activities they seem best suited to perform. Policy reforms aimed at achieving this, together with reforms to improve the performance of public sector seed activities, must be seriously considered if countries are to continue to make progress in agricultural development. Past approaches to seed policy appear to have been good enough to reach Stage 2 or, in some cases, Stage 3, but may not be adequate to move beyond these levels, particularly in the rain-fed agricultural regions of the developing world where small farmers predominate.

It has been argued above that improved seed is a critical component of agricultural development and that the challenges facing seed producers in developing countries are greater than ever. To meet these challenges, reform of existing national seed programs and policies is essential. A review in the next chapter of the experiences of several key developing countries should help to clarify these issues and suggest possible solutions.

FOOTNOTES, CHAPTER 6

1. World Bank, *World Development Report* (WDR), 1982, p. 50.
2. Ibid., p. 59.
3. Ibid., p. 61.
4. Ibid., p. 70, using yields in the Punjab as an example; also discussions with agricultural experts.
5. Federico Poey, "Characteristics of Seed Saved by the Small Farmer," paper presented to CIAT Seed Conference, Cali, Colombia, October 1982.
6. Johnson Douglas, *Successful Seed Programs, A Planning and Management Guide* (Boulder, Colo.: Westview Press, 1980), p. 23. This is an excellent presentation on setting up a seed program in the developing country context.

Seed Policies in Several Developing Countries 7

COUNTRY STUDIES

Information is fragmentary on seed programs and policies in developing countries. There is not yet a book or collection of studies that reviews the experiences of several countries in a consistent or complete way. This chapter does not fill that gap, but rather is an attempt to distill information from papers, articles and reports on the progress of the seed industry in some developing countries. The objective is to examine several situations to give the reader a general idea of the successes and failures that have occurred in recent decades. Clearly, more work needs to be done in this area, and it is to be hoped that one of the international organizations with broad experience in agricultural seed programs would undertake to prepare and publish a series of frank and factual reviews of countries' experiences.

The countries to be discussed here are India, Thailand, Turkey, Kenya, Brazil, Mexico, and Guatemala. In part this selection is based on the availability of information, but more to the point is the importance of India, Brazil and Mexico in the overall setting of agriculture in the Third World. The other countries present interesting case studies of seed programs in diverse settings.

India

Until 1960, agriculture in India was overwhelmingly characterized by Stage 1 of Johnson Douglas's taxonomy, outlined in the previous chapter, with traditional agricultural practices and seed saved from the previous year's crop. Breeding programs existed, but had not made much impact. Beginning in the early 1960s, genetic materials based on semidwarf wheat and rice varieties and hybrid corn varieties were introduced into Indian breeding programs. The potential for yield gains was immediately apparent, and the government of India established the National Seed Corporation (NSC) in 1963. By mid-decade, hybrid and high yielding varieties had been introduced in demonstration programs for several crops and, based on the results, demand for improved seed grew rapidly. Seed multiplication was a government specialty and was expanded rapidly to accelerate the availability of new varieties. The blue tag on this seed quickly became a mark of quality.

Some Indian states set up their own semiautonomous public sector seed companies and, in principle at least, private companies were encouraged. Tarai Development Corporation of Uttar Pradesh state was a model for the state seed agencies. Tarai used contracting farmers to grow seed, and it institutionalized certification and quality control. Revenues from sales were channeled back to research, which was used to develop new strains of high yielding wheat and rice that possessed better resistance characteristics than the initial high yielding varieties. In the region served, certified seed went from 5 percent to 40 percent of demand in just a few years.

Problems, however, began to develop almost immediately. Although certification of seed was provided for in 1966, a separate agency had not been set up for this. Complaints about seed quality surfaced as early as 1967, and in 1971 the National Commission of Agriculture reported that the quality of high yielding varieties was "deteriorating rapidly."[1] The high yielding varieties were in short supply relative to demand, and other types of improved seed were not available. The other semiautonomous state companies were not as successful as Tarai, and there was overexpansion of this concept as proposals for a company in each state were advanced. This led to declining seed quality and to financial losses.[2]

In practice, private companies were not encouraged; rather they were discriminated against. They had to fit into the niches of the seed industry that the government did not reserve for itself. As a result, the private companies were hostage to changes in government policy, to competition with subsidized state seed companies, and to deteriorating quality control in the seed supplied to them by the public sector multiplication programs.

By the early 1970s, "companies began to have serious misgivings about their future in the Indian seed industry."[3] By the mid-1970s, when the World Bank was preparing its first seed project in India, one of the underlying policies was to provide a basis for the growth of the private sector and to improve the balance between public and private components of the seed industry.[4] The objective was to encourage competition and protect the private seed companies from discrimination.

The World Bank seed project also aimed to strengthen the semiautonomous state seed companies and to structure them along commercial lines. Farmers who would contract the seed out would own shares, as would the state governments and the NSC. The state seed companies would buy foundation (basic) seed from the NSC and sell it to the growers. The NSC was also the channel for assistance from the World Bank and the International Agricultural Research Centers. The project was reasonably well designed but, in a rush to move ahead, preparation and implementation were not wholly adequate. With regard to the Haryana state company, "no demand survey could be made; training in seed production could not be given to personnel; needed facilities for processing (such as drying) could not be provided; and proper storage for seed could not be arranged."[5] There was a notable weakness in staffing; out of 125 established staff and officers, none was a trained manager and only one had attended a short training program in marketing. The program's incentives were poorly structured, and insufficient technical assistance was provided to the contracting farmers. These farmers took the heaviest risks and were reluctant participants. Their production had to be certified to be accepted for payment, and there was some feeling that this process was corrupted. In some cases, damage by rain and flood hurt production, with the contracting farmer taking the brunt of the loss.

On the demand side, local farmers were often unwilling to buy the certified seed at 31 percent higher than the value of their own seed. The problem was not just price, but also seed quality, performance and availability. To raise sales, a subsidy was introduced, and the improved seed sold at 15 to 20 percent above the grain price. The new seed was expected to yield 10 percent increases, so it was hoped that over time the subsidy could be removed.[6] Within two to three years, demand did pick up, the subsidy declined and demand began to outrun production. Production targets were not met largely because of management problems, lack of incentives for contract growers, and vagaries in the weather. It was found that the farmers using the seed were disproportionately land owners and high caste farmers; therefore the seed program did not improve income distribution. And many did not know about the proper use of irrigation, fertilizer or pesticides.[7]

These are the types of problems and difficulties encountered in the building of a seed industry in a country that has made significant advances during the last 20 years. There is no doubt that progress has been achieved, but it is also clear that much remains to be done. In its *World Development Report* of 1982, the World Bank presents a very positive view of the post-1975 evolution of the private sector component of the Indian seed industry.[8] Some private companies were allowed to produce foundation seed at that time, resulting in better availability of high quality seed for farmers as well as faster introduction of new varieties. A network of more than 10,000 seed dealers has spread throughout the country, and farmers have developed preferences for brands with proven performance. Profits have been plowed back into genetic research on crops such as sorghum, millet, cotton, corn, sunflower, and certain vegetables. The WDR section on India concludes that this "... is an important example of constructive, competitive interaction between the public and private sectors."[9]

Most observers would agree that the situation has improved since 1975, but would not be as positive as the *World Development Report*. For private companies conditions are better, but they are still discriminated against. They must fit their programs into areas that the government has not monopolized for the NSC or the semiautonomous state agencies. Wheat and rice are the leading crops, but are not among the crops mentioned above; research in those crops is done in the public sector using basic research conducted by the IARCs. Once a company has established itself and built up a market, there is no guarantee that the public sector will not change the rules, move in with the large resources of the public companies and, with subsidized sales, take over the market that the private company developed. Most important, private companies continue to face legal obstacles and regulatory disadvantages. The variety trials leading to seed certification, for example, are still conducted by the NSC, the public sector institution, thus creating a strong potential for conflict of interest. This situation persists despite recommendations dating back to at least 1971 that the official trials should be conducted by a knowledgeable and impartial agency. Such an institutional reform would help to make the trials fairer and to encourage the approval of the new varieties with the best performance characteristics.

In general, private breeders face excessive legislation and bureaucratic inertia. Variety approval and seed certification are becoming more and more important to market success, as they should, but in many cases the trials are poorly conducted due to insufficient funds, inadequate facilities and ineffective staff. Under these circumstances, public breeders' varieties receive priority attention, and private varieties are frequently held up. Finally, the quarantine and other restrictions on imports of seed tend to favor the public sector. Imports have to pass through government channels that are connected to the NSC. Private companies with overseas connections are understandably reluctant to let proprietary varieties, particularly elite breeding lines, run the risks inherent in these procedures. The result is that India is deprived of some of the most advanced genetic material in the world. Quarantine restrictions are certainly needed, but must be carefully structured and monitored if the country is to have access to the widest spectrum of worldwide genetic resources.

In sum, the "constructive, competitive interaction" that has evolved since 1975 has resulted from a reduction in—not an end to—discrimination against private companies. And further reductions can result in considerable additional improvement in both the public and private components of the Indian seed industry. The public sector seed agencies need substantial upgrading in terms of management, training, proper incentives for staff and contract growers, and facilities. However, more resources channeled to the public sector could be self-defeating if legal and institutional changes do not achieve more equal treatment of private seed organizations vis à vis their public sector competitors. Without a better internal incentive structure and more external competition, the NSC and the semiautonomous state companies are not likely to improve their efficiency significantly. The Indian seed industry has made important strides in moving from Stage 1 in 1960 to a mix of Stage 2 and Stage 3 in 1980. The benefits to Indian agriculture and the Indian people have been enormous. Efforts now must focus on reaching a mix of Stage 3 and Stage 4 by the year 2000.

Thailand

In terms of the development of its agricultural sector and seed industry, Thailand is an example of Stage 2 moving into Stage 3. Two fundamental differences stand out in comparison with India. First, Thailand is a land-extensive country where cropland has expanded at 4 percent per annum since 1960, and agricultural products are consistently exported. Second, the public sector is less dominant in the seed industry and has a stated policy of encouraging the private sector to take over seed multiplication and distribution wherever it can effectively do so. The Royal Thai Government's position is that it will undertake the projects in the seed industry that the private sector does not. As an example, the public sector will not produce hybrid seed and will adjust its production of open-pollinated varieties depending upon the output pattern of the private sector.[10]

Thailand's agricultural development strategy, as set out in the 1982–86 five-year plan, emphasizes increased yields rather than a continuation of the previous pattern of stagnating yields and rapid growth of cultivated land. The basis of the plan is to increase inputs of agricultural chemicals, irrigation, extension services, and improved seed; fertilizer use, for instance, is very low. There is enormous potential for greater yields if the farmers can be shown how to raise their income through greater use of these inputs.

Rice is the most important crop in Thailand, comprising between 35 and 40 percent of total crop value, but corn, sorghum, soybeans, mung beans, cassava, and sugar cane have experienced strong growth in recent years. Demand for Thai products has been good, and the export promotion of corn and sorghum has been particularly successful. Thailand, therefore, sees increased agricultural output as a foreign exchange earner in world markets.

Demand for improved seed is quite strong in Thailand, exceeding supply, and continued expansion of improved seed supplies is a key part of the government's agricultural development strategy. Beginning in 1976, the Agency for International Development (AID) supported seed projects aimed at greater overall use of improved seed and strengthening the private sector's capabilities. The leading farmers seem to understand the value of seed with proven performance and are willing to travel considerable distances and buy well in advance to make certain they obtain varieties that combine high yield with resistance to local pests and diseases. These varieties are in exceptionally strong demand, and the potential market for improved seed is substantial. To encourage private companies, special incentives for investment in the seed sector have been established. The crops of most interest to private seed companies are corn, sorghum, soybeans, groundnuts, and various vegetables. Rice seed is less attractive because of government subsidy programs that hold down its price.

In recent years, production of improved seed has grown rapidly, and Thailand is now at a point where some consolidation and rationalization may be needed before it can confidently move into Stage 3 of its seed industry development. Several problems have emerged that must be addressed:

(1) slow and inefficient decisionmaking within the public sector seed organizations;

(2) limited effectiveness of agricultural research due to shortages of trained manpower, over-centralization and poor linkages with the extension system;

(3) poor supervision and provision of technical assistance to contract growers;

(4) inadequate incentives for contract growers given the quality standards they must meet;

(5) weak distribution and marketing that must be improved if seed is ultimately to reach the great majority of farmers on a timely basis;

(6) low seed prices that have resulted in a deterioration in the financial position of public sector producers;

(7) lack of a seed certification law and an objective agency to conduct the necessary trials;

(8) cumbersome and uneven regulations.

There is a central contradiction in Thailand's seed policies that must be resolved before significant additional progress can be made. There seems to be an unwillingness to let price act as an equilibrating factor in the seed market, particularly for rice. Prices are held down and, as a consequence, demand exceeds supply, the public sector agencies need subsidies to shore up their losses, and marketing and distribution systems are underdeveloped. Until prices are allowed to rise to an appropriate level, all of these problems will persist. No effort or expertise in the distribution aspect is needed because there is excess demand for whatever seed can be produced. The most enterprising farmers find some way to obtain the improved seed, but many others are unable to or are not even aware of the seed. An established seed distribution network is needed that will reach the villages and bring improved seed to all who can use it. Yet such a system is unlikely to emerge as long as demand exceeds supply at the prevailing price.

A derivative set of consequences of low prices is that, despite tax incentives and various blandishments, private seed companies will not commit to serve a market unless a strong likelihood of profits is visible in the years ahead. Private investors must have a return on their investment before tax incentives have any meaning. Low prices also reduce resources available for research and the development of new varieties. Private research must be financed out of seed sales, and public research must be funded by either tax revenues or product sales. The price issue cannot be avoided and is fundamental to solving the most important problems facing the Thai seed program.

As a general proposition, it appears that low prices are a useful device for moving from Stage 1 to Stage 2 in the development of a national seed industry, as the demand for improved seed rises rapidly. However, low prices can then become an obstacle to progressing beyond Stage 2. If prices are not high enough to cover costs fully, public sector production must be subsidized, the private sector cannot emerge as a real contributor, the public sector seed agencies tend to become progressively more bureaucratic and inefficient over time, and an effective, mature seed industry never emerges. Persistently low seed prices and heavy dependence on public sector seed production can have serious consequences for agriculture in developing countries. Countries may become permanently stuck between Stage 2 and Stage 3, with the domestic seed industry unable to play its proper role in agricultural development.

Thailand is a long way from such an outcome, but persistence in underpricing improved seed could make the transition to Stage 3 unnecessarily difficult and time consuming. With demand for improved seed so strong, the temptation is great to expand production too rapidly. This in turn can lead to declining quality and to some of the problems experienced by India in the early 1970s. Thailand must confront its problems directly before significant further expansion can be undertaken. Emphasis on development of a vigorous private seed sector is correct, but is unlikely to work unless the pricing problem is successfully resolved.

In a review of World Bank experience in Asian seed projects, including the Tarai seed project in India, 10 important lessons were noted,[11] some of which have been included in the discussions here on India and Thailand:

    (1) success is more likely when building onto an existing successful program than when starting from scratch;

    (2) success is more likely with flexible and dynamic management than with a public sector agency having restricted autonomy;

    (3) large capital intensive projects are often inappropriate;

(4) public sector agencies are typically less efficient than private farmers (at tasks that both can do);

(5) farmers are more involved and effective when their share in the profit is assured;

(6) public sector agencies have been slow to develop the capability of producing good quality breeder and foundation seed;

(7) quality control and certification by an impartial agency is indispensable for success;

(8) social amenities must be adequate in project areas if good staff is to be retained;

(9) trainees must have attractive incentives if they are to stay with the project;

(10) projects tend to overestimate demand for open-pollinated seed such as rice or wheat.

In going beyond these 10 points, the World Bank review observed that farmers have sometimes been supplied with inappropriate seed, which has tended to give "government seed" a bad name. In more than one country, only small amounts of the appropriate varieties of improved seed reach the farmer in a timely way, even though large sums of government and donor financing are going into the seed program. Finally, the review noted the great need to improve the linkage between research and the farmer through a more effective extension service.

Kenya

Kenya is an interesting case of a country where the seed industry started with a private sector organization, which has since become predominantly publicly owned. Founded in 1955 at Kitale by a small group of local farmers, the Kenya Seed Company was originally a producer of grass seed. Over the years, however, its success has been based primarily on hybrid corn seed introduced in the mid-1960s. Therefore, for certain crops Kenya has been in Stage 3 of seed industry development for 20 years. Although this pattern is atypical for both Africa and the Third World, it is instructive to consider the evolution of the KSC because its approach to certain problems may apply in other countries.

The KSC was set up to utilize basic research conducted at the Kitale Grassland Research Station. In time, this research facility became the national Agricultural Research Station, as breeding work expanded to include corn and other crops in addition to grasses. Within the KSC, a research program was established with the objectives of maintaining breeder seed for corn and sunflower hybrids, breeding and selecting new varieties of corn and sunflowers, and testing responses to pests, diseases and chemical applications. Basic seed was produced on irrigated company land with separate handling and conditioning facilities, and multiplication was done through contract growers in the Kitale region. The KSC provides a management service for farmers who have land suitable for seed multiplication, but who lack financial or managerial resources. The KSC has its own selection procedures to ensure quality, but official certification is done by the National Seed Quality Control Service attached to the Research Division of the Ministry of Agriculture.[12] Throughout the 1970s, the government of Kenya maintained a pragmatic attitude toward regulation; to help the KSC, the government made a point of avoiding the imposition of overly stringent regulations. Technicians were brought in from abroad, particularly the Netherlands, and Kenyans were trained.

The KSC has generally been successful in overcoming the many problems it has faced over the years. Marketing and distribution have been particularly difficult in reaching small farmers. W.H. Verburgt, Managing Director of the KSC, relates his experiences:

> Distribution seems to be the bottleneck of most [seed] organizations no matter how they are organized or owned. Distribution is particularly important to top management in order to reach small farmers within a very large and wide area. This requires attention; it is no easy task. But it is possible if you work hard and do not try to do it over a short span of time. In Kenya it took us at least 10 years, and I will admit that this was longer than we expected when we began.
> Many problems have to be overcome. Storage points have to be organized. Physical movement of seed needs a lot of attention, particularly in an area where there are hardly any roads. Furthermore, it is difficult to make sure that adequate quantities of seed are available at the right time and in the right place.[13]

In the early years, demand for improved seed outran supplies, as is characteristic of Stage 2 in seed industry development. The KSC expanded output as fast as was consistent with maintaining quality, but did not reach a balance, with some carryover stock, until the late 1960s. Its attitude was that the best farmers would likely obtain their seed early, which resulted in good yields for KSC products, increasing demand correspondingly the next year. The KSC aims to have a couple of distributors in each district who also try to act as extension agents linking seed research and the farmer. To simplify operations, all sales are on a cash basis, and seed sells for the same price throughout the country. It is the philosophy of the KSC that "the marketing of seed should start with the plant breeder."[14] Breeders must look at the farmers' position and take into account the effect of new research on the farmers' net income rather than just the level of yield. In Kenya, both company distributors and the government extension service cooperated in providing information to breeders on the realities facing farmers.

Pricing policies were an important part of the successful seed strategy of the KSC. "On the one hand, the price of the seed should enable the farmer to benefit substantially from his investment. On the other hand, the seed grower and the seed organization should get a fair return on their money as well."[15] It is a real challenge for management to produce high value added seed efficiently so that incentives are attractive for all involved. Contract growers are a critical part of seed production and must be compensated for the greater effort and risk they have in seed multiplication. The seed organization must cover research, overhead and operating costs, in addition to paying a return on capital and/or debt. Marketing and distribution in developing countries with poor transportation and communication systems entail high costs. In particular, the final distributors should get a good return for their efforts because they are the direct contact with the farmers and make all the other seed operations worthwhile.

In contrast to full cost pricing, subsidies tend to lead to several unfortunate results. In the short run, they can help lower the cost to the farmer while providing a good price to the seed producer. In the longer run, there is the issue of what happens when the subsidy is phased out. Farmers must pay more than they are accustomed to, while producers will be selling less seed and receiving lower revenues than before. As sales decline, unsold seed stocks will build up. Subsidies act as a barrier to the entry of new unsubsidized producers. Subsidies also tend to reduce incentives for the producer to serve the farmers and tend to increase time and effort spent dealing with the source of the subsidy. Over time, subsidies can result in cumulating inefficiencies and stagnation in the development of the seed industry. Subsidies are a short-term palliative that frequently creates difficult problems and acts as a constraint on progress in the long run. The success of the KSC suggests that a long-term perspective is desirable in developing a nation's seed industry. If initial expansion is slower, a solid foundation is established, seed quality is kept high, subsidies are avoided, and regulation is kept to a minimum. Progress from Stage 1 to Stage 2 may be slower, but progress after Stage 2 is likely to be faster.

By the mid-1970s, the KSC had produced 13,000 m.t. of corn, 8,000 m.t. of wheat, 3,000 m.t. of sunflower, and 11 m.t. of grass seed.[16] Corn is the dominant crop within Kenya's agricultural economy, providing the staple food in the diets of nearly 90 percent of the people. By the mid-1970s, hybrid seed was used on more than 40 percent of the one million hectares planted in corn.[17] This is particularly interesting because the vast majority of those using hybrid seed were small farmers. It has been argued that hybrids are too expensive and too difficult to cultivate for extensive use by small farmers.

The International Agricultural Research Centers led by CIMMYT, the corn and wheat center, feel that improved open-pollinated varieties offer the best prospects for corn cultivation in the developing countries. They have identified five criteria for selecting between hybrids and open-pollinated seed:

     (1) environmental limitations on yield;
     (2) infrastructure limits on markets for input and output;
     (3) the financial strength and risk-taking capability of the farmer;
     (4) the resources available for seed research and production;
     (5) the time horizon for progress in seed development.[18]

The more positive these factors are, the more likely a hybrid program is to succeed. Generally, CIMMYT has taken the position that hybrid seed will not have an impact on yields in most parts of Africa and Asia at any time soon. Others, including the international hybrid corn seed companies, are more optimistic. The upshot of this is that CIMMYT research is concentrating on open-pollinated varieties, while the private companies are working on hybrids. The outcome seems to be healthy competition, with farmers benefiting from a wide choice of potentially suitable varieties and agricultural practices.

In Kenya, hybrid corn was introduced to the farmer as a separate crop and combined in a package with improved agricultural techniques and inputs. There was complete coordination and cooperation among the important organizations in Kenyan agriculture: the Ministry of Agriculture, the Farmers' Association, the KSC, and the other suppliers of inputs. The seed was advertised as yielding 20 to 30 percent more, but some farmers claimed gains of 100 percent. Using a combined approach, it was possible to make a big impact even though the customers were small farmers. By the early 1980s, 90 percent of the hybrid seed used in Kenya was purchased by small farmers in 10 kilogram units (0.4 hectare) or 25 kilogram units (1.0 hectare). In Kenya, hybrid seed has been well accepted by small farmers operating under conditions that are not very favorable in terms of the five criteria noted above.

The Kenyan seed experience is unique and cannot be replicated, but there are some useful lessons for seed development elsewhere. Probably the most important of these relate to emphasizing a proper pricing strategy and an adequate distribution system. Among the issues that remain for the seed industry in Kenya are the implications of the nationalization of the KSC and the need for additional seed producers to improve competition within the country. Over the years, the private shareholders who started the KSC have sold their shares, and the Kenyan government has become the majority owner in stages. It is not clear what effect, if any, this will have on the performance of the KSC or on the development of the seed industry in Kenya. Only time will tell if KSC will continue its successes and whether several other seed organizations will emerge to create a strong, diversified seed industry.'

Turkey

Through the 1960s and 1970s, Turkey de-emphasized agriculture in its development strategy and focused on an import substitution approach to industrialization. Agriculture's role was to provide low cost food for the urban work force and raw materials for use as industrial inputs. To achieve this, policy relied on considerable public sector support of agriculture through subsidies for output prices, credit and input costs. The result was deteriorating performance by the agricultural sector; during the late 1970s and early 1980s, growth of output averaged only about a half a percent per annum.[19]

By 1980, the Turkish economy was in deep crisis. It was clear that Turkey's development strategies were not producing results in agriculture or industry, and policies were realigned across the board to improve incentives and reduce market distortion. In agriculture, improvement was already evident by the mid-1980s, with a long way still to go, particularly in seeds. Some estimate that only about 5 percent of seed needs are being met by the seed industry and that use of improved seed could raise yields by as much as 30 percent.[20] Turkish agricultural strategy has focused on four priority areas: livestock, poultry, fruits and vegetables, and seed. Agriculture is felt to be an area of comparative advantage for Turkey, and the goal is to increase exports of food products to Western Europe, Eastern Europe and the food deficit countries in the Middle East. To achieve this goal, Turkey must have a major turnaround in its seed industry to generate yield increases in its major crops—wheat, vegetables, sugar beets, and barley. Corn and soybeans are still minor crops, but should become increasingly important as feed for animals.

The Turkish seed industry is characterized by two interrelated problems: excess regulation and a debilitating proliferation of government agencies. Public sector seed activities are not coordinated, and by 1981 there were 28 separate institutes, research stations, agencies, universities, and secondary schools producing public sector seed for sale. Private activity is weak and tends to serve small specialized markets, i.e., tomatoes and vegetables such as sugar beets, forage and hybrid corn seed. The Turkish seed industry appears to be in Stage 1 approaching Stage 2 for most crops, with progress to Stage 3 in only a few crops and regions. Overwhelmingly the seed used by the farmer is saved from the previous year's crop, and indications are that this is causing declining yields, i.e., certain crops are experiencing genetic deterioration.

Prior to 1982, when the regulatory conditions were eased somewhat, the laws appeared to restrict seed available to public varieties, to keep the seed industry independent from foreign seed, and to maintain prices as low as was consistent with regulatory standards. The

results were similar to what had occurred elsewhere in these conditions:
(1) improved seed was often unavailable as demand exceeded supply;
(2) a black market sprang up for high value vegetable seed;
(3) constraints on seed supplies reduced yield gains;
(4) low prices and poor margins weakened the existing private firms and made foreign firms reluctant to invest in Turkey;
(5) subsidies were required to support public seed production;
(6) the choice of which varieties to introduce was made by breeders, not by farmers;
(7) frequently neither farmers, distributors nor extension agents knew what would be available for next year's crop;
(8) varieties available were often not relevant to farmers' needs and/or in the wrong quantity;
(9) the testing process tended to overemphasize yield under optimum conditions and did not adequately test for pest, disease or drought resistance.

Because of these difficulties, almost all farmers, both traditional and commercial, were forced to save their own seed to be sure of having something to plant the next year. Although the current distribution system is not a major bottleneck, once production begins to rise it is likely to become critical. Approximately 15 percent of public sector seed is sold by the producing institute or agency to local farmers. The rest is distributed through the Turkish Agricultural Supply Organization and, when credit is involved, through the Agricultural Bank of Turkey. The private sector works through a network of dealers, but this tends to be region- and crop-specific. The tying of credit to seed can lead to unfortunate outcomes when seed supplies are short or inappropriate. In some cases, farmers must accept whatever seed is available to get credit, and the seed may not be what the farmer wants or what the extension agent has recommended. The distribution system, while capable of supplying perhaps 5 percent of Turkish farmers, is probably not adequate to supply 30 or 40 percent of total requirements. Therefore, a strengthening of this part of the system would have to go hand in hand with the expansion of production.

Recognizing the importance of a strong seed industry, the government of Turkey in 1982 reorganized the national seed program. It introduced several major reforms with the goals of:
(1) accelerating public sector decisionmaking;
(2) improving information flows on seed requirements and availabilities;
(3) coordinating seed production through regular meetings of joint public/private commodity working groups;

(4) reducing the regulatory obstacles to obtaining permission to produce new varieties.

The government plans to maintain control of the two key crops—wheat and cotton—but the private sector is intended to have primary responsibility for seed in other crops.

Reserving wheat and cotton for the public sector may lead to problems of the types discussed above; that is, pressures to subsidize will be strong, inefficiences will increase and production may expand, but only with a decline in quality. The inadequate distribution system of the past and poor linkages between the farmer and the breeding program are likely to persist.

The government seems to want greater private sector involvement and has introduced various incentives to attract private investment. Most important, in 1983 the price of seed produced by the private sector was deregulated. This is subject to the condition that sales prices be based on actual costs and be submitted to the Ministry of Agriculture and Forestry three months prior to planting time. Several seed firms appear interested in setting up new operations in Turkey.

The goals and policy reforms seem to be moving in the right direction to bring Turkey to Stage 2 for the bulk of its agricultural system and to prepare the way to Stage 3. If the present seed strategies can be maintained, or intensified, then positive results should be evident by the late 1980s. The gestation period for seed policy changes is not short, and holding to the current course may be the biggest challenge now facing Turkish seed policymakers.

Mexico

During the 1970s, Mexico experienced sharply rising demand for food based on rapid expansion in income from petroleum and on continued growth in population. During the decade, income increased at an annual rate of about 6 percent, while the population growth rate was roughly 3 percent. In the 1960s, Mexican agriculture had expanded substantially, but in the 1970s, output grew only slightly faster than population. The bulk of the demand effect of rising per capita income, therefore, fell on imports that rose over the decade by about one-third. This deterioration in the balance between domestic demand and supply accelerated in the early 1980s primarily due to prolonged and extensive drought. Mexico's largest crop is corn, taking about 40 percent of total cultivated land; beans and sorghum are the next most important crops, followed by wheat. Irrigated agriculture accounts for between 25 and 30 percent of total land and produces 40 to 45 percent of output. Corn and beans are predominantly rain-fed crops, while wheat and export vegetables are mainly irrigated crops.[21]

The rain-fed areas are the older, more traditional agricultural regions of Mexico dominated by the "ejidos," small landholders whose rights are defined and guaranteed by the Mexican constitution. The irrigated areas tend to be in more remote regions where traditional agriculture was not as strong and landholdings were somewhat larger. During the 1960s and 1970s, Mexican agricultural development focused on capital-intensive projects, most notably irrigation systems, in the more remote areas. The irrigation projects have led to use of improved seed and better agricultural practices, resulting in increased output. In the traditional regions the challenge is much greater and in some respects is exacerbated by the legal and institutional rigidities built into the "ejido" system. The decline in the growth rate of agricultural output in the 1970s was significantly due to a decrease in the availability of previously uncultivated areas suitable for capital-intensive irrigation projects, while the problems of rain-fed agriculture were still not adequately addressed.

After 1977, emphasis on rain-fed agriculture increased, but adverse weather patterns between 1978 and 1983, and international financial problems after 1982, resulted in continued difficulties. Recognizing the problems facing the small farmers in the rain-fed areas and anticipating the continued growth of petroleum revenues, the government introduced a new agricultural strategy in 1980. The core of this approach was to strengthen the public agricultural agencies and provide strong incentives for rain-fed crops. The result, in 1981, was a one-third increase in personnel in the Secretariat of Agriculture and a substantial rise in expenditures on subsidies, which reached more than 20 percent of total agricultural output. In light of the softening petroleum prices in 1982 and decidedly worse international financial prospects, agricultural expenditures at those levels could not be maintained. In 1982 and 1983, agricultural policies were reviewed and readjusted within the context of the need for national economic stabilization. The agricultural strategy that emerged sought to promote greater efficiency in all aspects of the food production, processing and distribution system. Strict priorities were set up and public expenditures were rationalized.[22] This is the eventful environment in which the Mexican seed industry has been operating in recent years.

While a few private firms predate it, the establishment of the National Commission for Corn in the early 1950s was the beginning of the public sector's involvement in the seed industry. In 1961, this agency was transformed into a much larger organization, the National Producer of Seeds, PRONASE, with responsibility for producing and distributing seed for all important crops. There is also a private sector comprising more than 30 companies and seed producing groups. By the late 1970s, the private sector had 55 to 60 percent of production (by volume), but in 1981 and 1982, with increased subsidies, production by PRONASE rose sharply, ultimately reaching 60 percent. With the reorganization of public agriculture and reduction of subsidies in 1983, output by PRONASE fell about one-third, and the private/public ratio was nearly restored. Between 1977 and 1981, total seed production doubled in volume terms, from 178,624 metric tons to 360,451 m.t., then declined to 345,329 m.t. in 1982 and fell again in 1983. This surge in production was largely due to PRONASE and created an unusually large buildup in carryover stocks; in 1982 and 1983, these were 30 to 40 percent of annual sales.[23] Much of this overproduction was sold as grain rather than seed, and the problems of the early 1980s led to some changes in the mid-1980s. A beginning was made at restructuring PRONASE and reorienting its focus to serve rain-fed agriculture. Most important, subsidies have been reduced sharply. Mexico appears to be at or close to Stage 3 in seed industry development in the irrigated regions and in Stage 2 in the rain-fed areas. The potential for a successful seed industry is clearly present, but Mexico's public sector needs to reinforce its efforts in the rain-fed areas to bring the entire country up to Stage 3. Many of the same types of problems identified in other countries are present in the Mexican seed industry. These include poor quality control, public sector subsidies with the customary resulting problems, and an unsatisfactory distribution system.

Several public sector agencies are involved in seed breeding, production, distribution, and financing. INIA, the National Institute for Agricultural Investigation (i.e., research), does basic research on variety development and the release of breeder seed. PRONASE multiplies the seed received from INIA, processes, conditions, then distributes seed for use by farmers. SNICS, the National Inspection and Certification Service, evaluates and certifies for release new varieties developed by INIA and private companies. New varieties must be evaluated for three consecutive years to be certified, but delays sometimes occur and the process can take five or more years. SNICS is also responsible for enforcing seed quality through field inspections and seed analysis. BANRURAL is the Rural Development Bank that finances contract growers multiplying seed for PRONASE and private producers. It also accepts up to 70 percent of PRONASE's output, provides financing based on this output, and acts as distributor and creditor to farmers buying seed. Mexico has one of the largest and most active public sector seed producing systems in the world, with an efficient existing infrastructure that should be able to satisfy production requirements. However, the focus of this system ought to be shifted to the rain-fed areas and restructured. Before turning to a more detailed consideration of the problems facing the Mexican seed industry, let us briefly consider the private sector.

The major private sector organizations in the seed industry include approximately 30 seed companies and several seed producer groups and credit unions. Three-quarters of the firms are entirely owned by Mexicans; the rest have U.S. partners, generally in minority positions. The private companies are heavily concentrated in the three states that lead in terms of irrigated cropland and in Jalisco, which has sufficient rainfall to make it the country's largest commercial corn producer. The companies produce hybrid corn, hybrid sorghum, safflower, and cotton seed. Research in the private sector emphasizes the testing and selection of foreign varieties suitable for Mexico. The prices charged by the private sector tend to be higher than those in the public sector for comparable seed. In the case of hybrid corn, the price of private varieties is double the public price, mainly because of the quality of seed and the superior yields.

Private sector production of seed rose from 100,000 metric tons in 1977 to 170,000 m.t. in 1980 and 175,000 m.t. in 1981. Then, in response to overproduction and heavy subsidization by the public sector in 1982, private production fell to less than 140,000 metric tons.[24] Thus, the net effect of public sector subsidies in the early 1980s was to undermine the private seed organizations and increase public stocks of unsold seed while supplying the farmer with approximately the same volume of improved seed that would have been developed in the normal pattern of public and private seed production.

At the national level, there is no clear seed industry policy regarding how the public and private sectors should productively interrelate. The public sector is subsidized, which reduces the potential for genuine competition leading to greater sector efficiency; and planning and coordination of private and public sector operations are also absent, which results in, for instance, a concentration of both public and private resources on breeding, production and distribution in the irrigated regions while the rain-fed regions are given less emphasis. If the private sector has the capacity to provide seed to irrigated agriculture, then public sector resources — particularly subsidies — should be directed to serve the needs of farmers in rain-fed regions. Such a reallocation of effort would save the public sector money and expand the availability and use of improved seed nationally.

During the early 1980s, PRONASE attempted to expand public sector production and sale of improved seed as part of an overall strategy to raise food production in Mexico. As noted, this attempt was not successful, but it served to highlight many of the critical problems plaguing the Mexican seed industry.

(1) A general problem has been that increased levels of gross production have led to declining seed quality and an inability to market properly all that is produced.

(2) The quality control agency, SNICS, has too few qualified field staff members to inspect and analyze all the seed grown.

(3) SNICS has ties with other public sector seed agencies and

is not sufficiently rigorous in enforcing standards on public sector seed.

(4) Training in technical skills is needed in all of the seed agencies.

(5) Public research by INIA is oriented toward irrigated areas (60/40 ratio in terms of experiments), while about 70 percent of cultivated land is in the rain-fed areas.

(6) INIA does not adequately label and differentiate seed varieties regarding their suitability for irrigated or rain-fed applications.

(7) INIA does little or no research on seed germination, on viability over time or on storage characteristics.

(8) After a variety has been released by INIA, that variety is left on the certified list and generally not re-evaluated.

(9) INIA does not systematically maintain seed after it is initially released and does not provide PRONASE with new breeder seed each year; this raises questions concerning seed purity and the stability of the characteristics of the varieties.

(10) Even a cursory review of PRONASE operations raises questions concerning the expansion of facilities in areas well served by the private sector; this in turn leads to high transport costs to move unsold public seed to areas where PRONASE facilities remain uncompleted.

(11) There seems to be a mismatch between distribution facilities and markets and, as emphasis on rain-fed areas increases, this problem is likely to worsen in the years ahead.

(12) In the early 1980s, the public sector seemed inclined to use increased subsidies to reduce the role of the private seed sector in the irrigated area rather than to expand public sector activity in the rain-fed regions where the private sector was absent.

(13) Public sector seed production was heavily subsidized until 1983, and this resulted in many of the problems that typically stem from such policies, e.g., inefficiency in seed production operations, failure to produce what the farmers want, inadequate marketing and distribution systems, and the creation of a large, costly bureaucracy.

(14) Linkages between the farmer and both the research agency INIA and the producing agency PRONASE are almost entirely lacking in rain-fed areas and are weak in others.

(15) The government restricts seed imports without regard for quality differences or farmers' preferences, which reduces competitive pressure on domestic producers.

(16) Concerning agricultural credit, BANRURAL functions adequately, but in some areas may favor growers and users of

public sector seed and discriminate against private companies. (17) Finally, there is lack of a group for overall coordination of the Mexican seed industry; such a body would have to include adequate representation from the private sector to be effective.

Among the developing countries, Mexico has one of the older and most advanced seed industries. The physical facilities, for the most part, are adequate, the institutions are largely established, a core of trained staff is in place, and there is a well organized and efficient private sector. Many of the 17 problems listed above are being dealt with in the reorientation of agricultural strategy in Mexico, and it is hoped that the others will be dealt with in the near future. As with Turkey, however, the biggest question is whether the more realistic and correct policy course that was set in place will be followed long enough for it to produce results.

The primary goals for the public sector are to redirect its efforts to the rain-fed regions, to improve efficiency, to save public expenditures by letting the private seed sector do as much as possible, and to hold to these policies over time. The guiding principles in public sector seed policy should be the phasing out of subsidies in the irrigated areas and the redirection of efforts to help poorer farmers in the rain-fed regions. If the public sector can compete with the private sector on an equal footing (i.e., without subsidies) in the irrigated regions as well as in the rain-fed regions, the seed industry will be better off. However, if the public sector cannot compete without subsidies, then those regions and markets should be left to competition among private organizations. Given the fiscal limitations facing the government and the demonstrated competence of the private sector in the irrrigated regions, the public sector should focus its seed resources — particularly any subsidies — in the rain-fed regions. Such a strategy, if adhered to, would accelerate the development of the Mexican seed industry into a mixed pattern of Stage 3 and Stage 4 within 10 years.

Brazil

Agriculture in Brazil has performed very strongly over the past three decades. The country has many regions and climates and consequently many important crops, ranging from traditional mainstays such as sugar and coffee to relative newcomers like soybeans and oranges for juice concentrate. The level of development of the seed industry in Brazil reflects the disparities among the regions. In the northeast, dominated by small landholders, agriculture is beset by chronic problems, including a scarcity of good land relative to population, shortage of capital, insufficient adapted varieties, inadequate infrastructure, and a drought-prone climate. In this area the seed industry is in Stages 1 and 2 of development. In most areas of the more advanced agricultural regions in the south and southeast, the seed industry has progressed to Stage 3. A comparison of facilities of the state agency for producing basic seed showed that, in 1980, about 70 percent of capacity was in the south or southeastern regions, while only 15 percent was in the northeast. The activities of the private sector are even more heavily weighted to the south and southeast, resulting in the overall balance of the seed industry being heavily weighted against the northeast. Data for 1978–79 reflect this concentration by region and by crop (see Table 7–1). The more advanced level of improved seed usage is clearly evident for the south and southeast with the central region at an intermediate level. Among the various crops, black beans, a small farmer staple crop, are heavily reliant on farmer-saved seed in all regions.

During the 1950s and early 1960s, agriculture was not given much emphasis in Brazil's development strategy. Policies were particularly biased against export crops, and internal controls kept down the prices of staple crops. The driving force behind Brazilian agriculture during this period was the expansion of cultivated land. Between 1955 and 1965, production of domestic food crops rose at an annual rate of 5.7 percent, of which 4.4 percent was due to increased land.[25] Export and industrial crops (including cotton and sisal) grew at an annual rate of 9.4 percent, with 3.6 percent due to expansion of land area. Coffee experienced significant yield gains averaging 10.6 percent over the 10-year period. By the mid-1960s, Brazilian economic policies were beginning to change, and greater incentives were given to export oriented agriculture. Between 1966 and 1977, growth in domestic crops was

**TABLE 7-1. ESTIMATED USAGE OF IMPROVED SEED,
BRAZIL, 1978-79
(Percent of Total Area Planted)**

Crop	NE	SE	S	Central	Total
Beans (black)	4	2	8	1	4
Corn	4	70	33	40	40
Cotton	20	90	91	75	50
Rice	8	50	42	30	41
Soybeans	N.A.	67	90	40	70
Wheat	N.A.	67	100	40	90

N.A. = not available.

Source: National Secretariat for Prices and Supply, Ministry of Agriculture, Brazil.

down to an annual rate of 3.8 percent, of which 3.1 percent was based on expanded area. Export agriculture experienced remarkable expansion, growing at an average annual rate of almost 23 percent; land used for export crops rose at 4.8 percent, so most of the total increase was based on yields. Soybeans were the great success story of this period with output growing 37.6 percent a year and land under cultivation rising 30.9 percent annually. A good deal of land in the south that had previously been planted in coffee was shifted to soybean cultivation.

During this remarkable period, the Brazilian government used credit subsidies to encourage agricultural growth and adjust the composition of the crops produced. Farmers in selected lines of production, who could get access to public credit, faced a very favorable incentive system; to a considerable degree these have been the export oriented farmers located in the south and southeast. The growth figures during the 1966–77 period reflect the emphasis of the policymaker as domestic food production per capita grew by only 1 percent and yields rose by less than 1 percent.

Brazil has thus used its extensive land resources as the foundation of its strategy to supply domestic food needs. While in the past such a strategy was possible, even desirable, there is increasing concern that new land will be much more expensive to bring under cultivation. As the cost of clearing new land, building roads and providing other infrastructure increases, the option of raising yields on existing cropland becomes more attractive. Reflecting these trends, the Brazilian government expanded agricultural research in the late 1960s and 1970s. These efforts were aimed initially at the export crops in the southern and southeastern regions, but focus is now shifting to the northeast and to new regions in the north and central-western parts of the country.

In the late 1970s, several events combined to cause the government to place even more emphasis on agriculture. The second sharp increase in oil prices, the buildup of external debt, protectionist obstacles to manufactured exports, and two consecutive years of bad harvests in 1978 and 1979 came together to force a reevaluation of policies. The result was a new development plan for the 1980–85 period that called on agriculture to raise the supply of domestic foodstuffs, and thereby help control inflation; expand exports of soybeans, orange juice concentrates, cocoa, and other products; increase sugar cane production to convert to alcohol (replacing imported oil as a fuel for motor vehicles); and generate employment, helping to reduce urban and rural poverty. To achieve this set of objectives, agricultural credit subsidies were expanded and redirected somewhat to small farmers. In addition, price controls were removed or reduced, incentives for alcohol production were expanded, and greater efforts were to be made regarding infrastructure and research.

TABLE 7-2. AGRICULTURAL OUTPUT, BRAZIL
(Per Capita Basis*)

	Domestic Crops	Export Crops	Sugar Cane
1977	100.0	100.0	100.0
1978	82.1	86.3	105.1
1979	85.4	90.5	110.5
1980	94.8	113.7	115.6
1981	90.8	110.7	118.3
1982	97.9	103.7	138.5
1983	74.4	106.9	156.7

*Assumes population growth of 2.3 percent annually.

Source: Same as for Table 7-1.

By the mid-1980s, the results of this new plan in the agricultural sector were mixed at best. Considering production in three categories— domestic foodstuffs, export crops and sugar cane—it appears that only sugar cane production has been able to progress as desired. Table 7-2, showing agricultural output per capita over the 1977-83 period, reveals a sharp decline in domestic crops and a modest performance in export crops (3.3 percent annual growth). Not only did domestic food production fail to grow relative to population, it declined in absolute terms between 1977 and 1983. Many of the important crops had very poor harvests in 1983 due to drought in the northeast and flooding in the south. Coffee and cocoa were among the few strong performers. Output levels in 1984 were better for both domestic food and export crops; however, a severe drought in 1985-86 reversed these gains.

These unfortunate events emphasize the need to push ahead with policies that will help increase and stabilize yields. The top priority is to clear out the maze of subsidies and price controls, which have become so complicated and have been changed so often that everyone is confused. Farmers take several crop seasons to adjust to new incentives, and stability in these policies is critical if growth in output is to resume. On the structural side, more resources must go into improving the existing system of domestic food production rather than concentrating on infrastructure and research on export crops and letting producers of domestic food crops fend for themselves. Brazil can no longer rely almost entirely on the availability of cheap new land to generate increased domestic food supplies. In particular, programs aimed at finding solutions for the problems of small farmers in the semi-arid regions of the northeast need a high priority.

Although hybrid corn production by a private company began in 1945 and the government produced seed in the 1950s, Brazil's improved seed industry dates back only to the mid-1960s. In a short time it has made considerable strides. However, in comparison with the size of the country and the increasing importance of agriculture, the seed industry is small in size, uneven in coverage and greatly in need of expansion and strengthening. A national plan of seed improvement was drawn up in the late 1960s and implemented in the south and southeast in the 1970s. The result was the successful and rather rapid creation of an acceptably effective seed industry incorporating federal, state and private breeders and public, private and cooperative producers of seed. From the beginning, the Brazilian strategy called for working with private companies and local cooperatives. Federal breeders would release varieties to state agencies, approved private companies and cooperatives for multiplication and distribution. Despite some controls on the sales price of seed, this system worked well, and by the end of the 1970s the seed industry was well established in the south and southeast. Brazil avoided several of the most common problems of seed industry development in the initial stages.

(1) Subsidies were mainly on the demand side of the seed market, i.e., credit for farmers and incentives to produce crops, rather than on the seed production side.

(2) There was little or no discrimination against private seed producers and distributors.

(3) The federal government concentrated on research and has not yet developed a large bureaucracy of people in production and distribution.

(4) State government agencies and the extension services could distribute seed, but were not heavily subsidized in their activities and therefore had to be reasonably efficient to compete with private companies.

(5) Certain crops, e.g., hybrid corn, were left largely to private companies even for breeding.

(6) Overly rigorous certification and regulatory restrictions were avoided at the early stages of seed industry evolution.

Certainly the advanced state of agriculture in the south and southeast of Brazil had much to do with this success, but an appropriate strategy was also important.

Extensive use of subsidized credit was central to success in the 1970s, but with increased austerity in the 1980s, this has been cut back. As part of the agreement negotiated with the International Monetary Fund in 1983, agricultural credits to farmers in the central, south and southeast regions were reduced significantly. Offsetting this somewhat were increased relative prices for agricultural goods.[26] This implies a beginning in clearing up the distortions that had proliferated in recent years, and it will be interesting to see the reaction of agriculture and the seed industry in the affected regions. Farmers in the northeast who qualify should continue to get subsidized credit, but even this may be delayed. Subsidies on exported crops were cut and some shifts among crops, e.g., from soybeans to corn, are likely to occur if the changes are enduring. These shifts in policy will be a challenge to the seed industry, but prospects are good for successful adjustment.

In 1973, the Brazilian Agricultural Research Enterprise (EMBRAPA) was created to coordinate research with a similar agency (EMBRATER) to provide extension services. EMBRAPA includes the federal research centers and works closely with the state research centers. The most important of these is in the state of Sao Paulo, which has a very strong program and extensive experience. Sao Paulo established a seed certification program along international standards as early as 1968, whereas no equivalent program exists at the federal level even today.[27] EMBRAPA conducts the federal seed research program, produces genetic seed and encourages the production of basic seed. In 1976, EMBRAPA created a subsidiary, SPSB, to produce basic seed. Until then, only Sao Paulo among public centers produced basic seed that measured up to well defined minimum standards. The SPSB was to raise the output of basic seed while at least maintaining existing quality standards. Where possible, it would assist the private sector in multiplying seed for distribution and sale. And where necessary, it would promote and support the creation of new programs to produce seed in areas where private organizations had been unable to get started. Finally, the SPSB would have a role in training technicians and other seed personnel. The operating philosophy was to let the private sector do what it was capable of so that the public sector could concentrate its resources on the remaining areas. Over time, the goal of EMBRAPA is to move out of, as much as possible, production and distribution of commercial seed and to concentrate on research and the production of basic seed. This is the direction in which events seem to be moving in the south and southeast; in the northeast, however, the story is different.

Throughout the northeast, the seed industry has, for the most part, remained in Stage 1 or perhaps moved to the beginning of Stage 2. With a high proportion of small farmers and a difficult climate, progress has been very slow. The supply of improved seed is inadequate and unpredictable in availability. Unlike the situation in the south, the private sector—including both contract growers and companies to produce and distribute seed—has not played a strong role. These problems necessitated a change in strategy, and the SPSB has been forced to release its basic seed almost entirely to the state agricultural systems for multiplication. However, the states frequently do not follow up due to budget shortfalls or personnel losses, and in tight years they sell the basic seed for final use by farmers. As a result, supplies of improved seed can vary enormously depending on the year to year budget position of the state governments. There are also problems with storage, distribution and delivery of seed over the vast distances and difficult conditions that prevail in the region.

Another problem is that some "improved seed" from the south is occasionally brought north, and when it performs poorly, farmers become skeptical of the value of all improved seed. Therefore, given a limited and unpredictable supply of improved seed at relatively high prices, the response of farmers is overwhelmingly to stick with their own seed. Low yielding but locally adapted traditional varieties thus predominate throughout the northeast.

Until recently credit has not been available to the small farmer, and this precluded buying seed even when the farmer wanted to. The credit problem has improved somewhat through easier financing of seed purchases and an interesting buy-back program that the extension service provides. Farmers can obtain seed from the extension service by pledging to repay in kind with part of their crop. The exact amount of output returned to the extension service depends on the price of the crop and must equal in value the initial value of the improved seed. This system is easy to understand and implement and should be watched carefully to see how well it works over time.

A final problem in the northeast has been seed quality control. No common standard prevails across the country, as the SPSB has its set of standards and state programs have theirs. In the south and southeast this system seems to have been satisfactory. The SPSB has been able to find contracting farmers to grow basic seed while meeting quality standards. In the northeast, however, problems have developed for a variety of reasons:

(1) the erratic climate in the northeast makes seed production a high risk activity unless the land is irrigated;

(2) there are a limited number of farmers with suitable irrigated land;

(3) the individual plots tend to be small, requiring intensive field inspections;

(4) quality control is hard to maintain because of variation in the pre-basic seed and the farmers' failure to rogue diseased and off-types thoroughly.

As discussed above, variation in the quality of seed has undermined farmer acceptance, but it is hoped that this can be rectified with the current expansion of SPSB facilities described next.

In the early 1980s, as a part of the new agricultural development strategy, the emphasis in the Brazilian seed program began to shift to the northeast region and its particular problems. The seed plan was drawn up in 1980 and put into effect in late 1981.[28] Although all regions were covered, the northeast had top priority. Crops to be covered were beans, corn, cotton, soybeans, and rice. The approach was three-pronged and included establishment of an improved seed multiplication capability in each state; extensive credit and technical assistance for seed producers; and promotion of farmer use of improved seed through demonstration plots, seed producer cooperatives and strengthening the regional seed distribution network. The centerpiece of this program is SPSB's seed production complex at Petrolina along the banks of the Rio Sao Francisco in the state of Pernambuco. Here, in cooperation with the local irrigation project, the SPSB will carry out its own production of basic corn seed on 500 hectares of irrigated land. The Petrolina complex will result in an increase of regional seed production by roughly 25 percent during the five-year start-up period (1982 through 1986). Such an expansion, while important, can only scratch the surface of the seed requirements for the region as a whole. As shown in Table 7–1, improved seed is used on less than 10 percent of cultivated land in the northeast, so the potential for increased demand is very large.

At its other production complexes, the SPSB contracts out nearly all of the actual growing of the basic seed and then conducts the processing and conditioning of the seed. Thus, the Petrolina project will be carefully scrutinized to see how well the SPSB manages a seed growing operation. As noted above, previous efforts at using contract growers in the northeast have been unsatisfactory, and this new tack should be a step forward. On the other hand, it would be regrettable if success with the Petrolina experiment should open the door to extensive public sector activity in production and distribution as this route could lead to serious difficulties in the form of bureaucratic inertia, inefficiency, subsidies, and needless conflict with private seed producers. A great strength of the SPSB is that it has paid its own way since 1979 and showed an excess of receipts over expenses in 1980, 1981 and 1983. It does not fund research (that is provided by EM-BRAPA), but it at least does not divert public resources from research. The production manager of the SPSB has indicated that the organization intends to multiply and supply improved seed only until ". . . the time comes when the activity becomes profitable enough to get the private sector involved."[29] This principle, if adhered to, should ensure the continued success of the Brazilian seed industry.

In the south, seed certification programs based on international models are likely to be established for the most important crops within the next few years. In the past, an absence of debilitating regulation has been a strength of Brazilian seed policies. As the certification regulations are drafted, care will have to be taken to raise seed quality without weakening the production capabilities of the seed suppliers, both public and private. This will require the creation of a separate agency with an arm's length relationship with EMBRAPA and SPSB. The same standards and priorities should be applied to public and private breeders, and a system for holding the certification agency accountable for delays ought to be considered. The goal of the certification agency should be to facilitate the movement of new and better seed from the breeder to the farmer as quickly as possible. A positive step in this process would be for the government to consult with seed producers in designing the certification system.

Guatemala

While Guatemala is a small country with a seed industry that is not very advanced in many respects, it is an interesting case. It has established a program whereby the Institute of Agricultural Science and Technology conducts basic research, prepares and releases basic seed to private growers and encourages them to multiply and distribute the finished seed. The ICTA supervises the multiplication stages, provides some inputs, conditions, and stores the seed for a fee that covers costs. The grower is responsible for land, labor, promotion, distribution, and marketing the seed. Once the seed is released to the grower, ICTA never takes back possession; it is up to the farmer to utilize ICTA's technical services in managing and making a success of the process.[30]

In a country or region that is in Stages 1 and 2, with a generally weak private sector, this approach holds the promise of creating an efficient, albeit small-scale, seed industry. It is based on a cooperative, mutually reinforcing relationship between the public and private sectors. Although it is still too early to call this experiment a success and to view it as a model to be employed elsewhere, it is nonetheless an interesting new approach to seed industry development.

In 1981, agriculture was still a very significant part of the Guatemalan economy, producing 25 percent of gross domestic product, absorbing 54 percent of the labor force and providing 61 percent of exports. It is divided into two different types: lowland agriculture with larger tracts of land producing for domestic and foreign markets, and highland agriculture dominated by small farmers of Mayan Indian ancestry, producing principally for self-consumption. The various plains and valleys at differing altitudes give rise to several micro climates suitable for many tropical and temperate crops, including coffee, cotton, bananas, sugar, rice, corn, beans, wheat, and vegetables such as broccoli, cauliflower and snow peas.

Although the production of seed goes back to at least the 1950s, the Guatemalan seed industry has continued to be underdeveloped. Progress has been halting and intermittent at best, and as late as 1977 and 1978, most seed supplied to Guatemalan agriculture was imported, largely from neighboring El Salvador. Since 1980, these imports have decreased due to plant disease problems and other difficulties in El Salvador, and also due to expanding seed capacity in Guatemala.

In 1972, ICTA was created as a parastatal organization to advance science and technology in agriculture, responsible for all extension services, seed certification and public agricultural research. As outlined, ICTA was set up to work closely with farmers to derive realistic solutions to current problems and to get responses quickly from the farmers. The focus of ICTA was on staple crops, and research teams were established in both the lowland and highland regions. Seed production operations at ICTA began in 1975 and were based on the use of contract multipliers who were required to sell all their output back to ICTA. At this time, public sector agencies including ICTA were responsible for marketing, which did not go as well as anticipated. In both 1975 and 1976, about half of ICTA's total production remained unsold.

Facing this marketing dilemma, ICTA changed its strategy and agreed to purchase an unspecified share of the contract growers' output. ICTA also raised the suggested retail price of finished seed by 30 percent to encourage distributors to market more aggressively and to create incentives for the contract growers to become farmer-seedsmen. They were encouraged to sell to other farmers the finished seed not bought back by ICTA. Prices were raised again in 1978, which encouraged more seed producers to join the program utilizing the basic seed and services provided by ICTA. In 1978, ICTA began to phase out completely as a distributor and to turn this function over to the emerging farmer-seedsmen. If they wanted to market their seed under ICTA's name and logo, they must submit to ICTA supervision, quality controls and certification; if not, the seed would be marketed under private brand names. In any case, ICTA would provide the basic seed to anyone wanting it, whether for multiplication or breeding purposes.

While in theory all farmers are eligible to multiply and market seed, in practice it has been the larger farmers with better educational backgrounds and greater financial resources who have become involved. This is unfortunate from a redistributional perspective, but is probably desirable from the perspective of getting the program to work. Once the basic experiment is seen to be working and the initial lessons learned, then niches may be found or created for smaller farmers. One such possibility is as a contract grower for an established firm or farmer-seedsman. Another possibility is for ICTA to furnish a complete package of services for small growers as was done in Kenya. Small farmers should be able to become seedsmen in the highland regions because of the many local variations in climate and conditions, but they will need considerable technical support from ICTA. Finally, if small seed producers are to emerge, they will need adequate credit facilities provided by the public sector. ICTA might be able to help using some variation of the in-kind repayment scheme that has been tried in Brazil. The inadequacy of financial resources is probably the biggest hurdle facing small farmers who would like to become seedsmen.

The seed industry in Guatemala is clearly less advanced than the programs in Mexico, Brazil and many other countries; yet at an early stage, Guatemala is taking a novel and creative approach to overcoming critical weaknesses found in many more advanced programs. Large-scale production subsidies are being avoided, the public sector is concentrating on research and the provision of basic seed, and local farmers are being encouraged to enter the seed business and market and distribute the seed they produce. The experiment being conducted in Guatemala is thus important and deserves continuing attention and analysis.

It is important to note some potential problems in the Guatemalan seed strategy. One is the retention of seed certification responsibility by ICTA, which is also involved in research and basic seed production. This is potentially a conflict of interest and a temptation to favor those seed producers within the ICTA system. A second potential problem is the evolution of a closed system where a few larger seedsmen might develop control of marketing or get preferential treatment in basic seed distribution. Such a trend could be especially difficult to control if these seedsmen are able to reinforce their economic position through the political system. Such a pattern would be as anticompetitive as a public sector monopoly and probably as inefficient in terms of failing to meet farmers' needs.

PATTERNS OF SEED INDUSTRY DEVELOPMENT AND POLICY OBJECTIVES

Against the background of the experiences with seed programs and policies of the seven developing countries discussed above, let us re-examine the four stages of agricultural and seed development identified in the previous chapter. The first stage is characterized by traditional agriculture and farmer-saved seed. In moving from Stage 1 to Stage 2, the driving force is excess demand for new varieties. During this period, the built-in resistance of traditional farmers is progressively overcome as they realize what is possible with improved seed. Farmers' demand for improved seed exceeds supply; this demand and supply imbalance seems to be characteristic of Stage 2. As seed producers struggle to catch up and suppliers rush to expand, there is the risk that seed quality will fall. Emphasis within the seed industry is on production to meet demand, and insufficient attention may be given to quality control and distribution. Agricultural extension systems are also usually underdeveloped at this point, and the private sector typically consists of only a few firms with very limited research capabilities and production coverage of just a few crops.

Movement out of Stage 2 is marked by the quality and quantity of seed supplies beginning to catch up with demand for improved seed. The first response of the seed producers to excess demand in Stage 2 is a major effort at increased aggregate production. Without improvement in distribution, however, seed supplies may end up in warehouses and the end-of-the-year carryover stock can rise to 40 to 50 percent, as occurred in Mexico and Guatemala. The immediate obstacle is not bulk production, but producing the right seeds and getting them sold and distributed at the right time and place. Experience shows that Stage 3 is not actually reached until the distributional system begins to sort itself out. Increased output must be accompanied by a beginning in the solution of customary problems with seed quality control, the extension system, seed promotion and distribution, and an emerging viable private seed sector.

Movement from Stage 3 to Stage 4 signifies the evolution of a fully mature seed industry with well established links to the rest of the world. Research and development continue to improve and broaden, supply and demand grow in approximate balance, and seed quality is guaranteed by the effective enforcement of certification laws. The greatest changes, however, are in the continued development of the seed multiplication, conditioning and distribution operations and the corollary growth of the private sector. Just as the evolution of a living organism is characterized by greater complexity and specialization of its constituent systems, so is the development of a seed industry: a healthy seed industry at an advanced level of development appears to be characterized by increasing complexity and specialization. In particular, the public sector moves in the direction of research and development, administering seed laws, educating needed technical specialists, and providing extension services. The private sector focuses on multiplication and distribution where competition and the incentive structure of the market forces them to respond to farmers' needs. When returns on research expenditures can be captured in sales, as with hybrids for example, private companies can also effectively undertake breeding activities. When the gains from research cannot be captured through increased sales, as with many open-pollinated crops, it is appropriate for the public sector to provide basic seed to contract growers and private organizations.

The underlying factor throughout these four stages is plant research and development. However, while the introduction of new improved varieties is necessary for seed development, it alone is not sufficient. It has been shown that the process is launched by the initial creation of a disequilibrium, i.e., excess demand for improved seed, which propels the system toward Stage 2. Progress, in the form of reaching Stage 2 and moving on to Stages 3 and 4, is dependent upon various components of the seed industry—production, quality control, distribution, and the private sector—responding to the disequilibrium with a series of constructive and mutually reinforcing adjustments. If the appropriate response of one of the key components is inhibited by inappropriate policies, the development process will be retarded. And, of course, it is possible for policies to be so inappropriate that countries regress in terms of the development of their seed industry.

In the concluding remarks of this chapter, emphasis will be on identifying those policy objectives that tend to encourage the desired response from the key seed industry components. Three basic perspectives will guide this discussion:

- the view of the development process as an interrelated whole with a consequent need to avoid policies that seem to be shortcuts but which in fact exacerbate longer-term problems;
- the need to work toward an allocation of functions between the private and public seed sectors that experience has shown to be most beneficial for the overall agricultural development of the country;
- the need for a clear and consistent seed development strategy that simultaneously guides policy decisions while keeping seed laws and regulations to a minimum.

While specific policy objectives will have varying relevance in different situations, these three perspectives are generally relevant at all stages of seed industry development across a broad range of countries.

In most cases, the breeding segment of seed operations is one of the strongest parts of the program and should not be a high priority for radical restructuring. The breeding problems needing attention center on facilities, staff, more awareness of farmers' specific needs, and access to a broader spectrum of genetic resources. Facilities have to be expanded and upgraded on a continuing basis if seed supplies are to keep up with demand. An even more critical need is for an expanded pool of people trained in breeding techniques, for additional training of those presently in breeding programs, and for better compensation of key people in those programs. Some restructuring seems called for in forging stronger linkages with farmers, especially small farmers, and in research directed toward farmers' needs. Examples of some of these needs were described in the previous chapter and include yield stability over a range of conditions, particularly moisture shortfalls, grain qualities and appearance, cropping practices, and plant structure. Finally, national breeding programs need to broaden the genetic base of their breeding materials. In most cases the appropriate materials are available through the IARC system or from public agencies in other countries, but with a shortage of staff and training, it is not possible for breeders in developing countries to follow up on even a small portion of the freely available germplasm resources.

The demand side of the picture will be considered first in discussing policies and objectives for seed development, especially moving from Stage 1 to Stage 2, when the driving force is excess demand. Among the small farmers in particular, the great need is for programs to increase demand for improved seed. The first requirement for stimulating demand is the availability of superior varieties of locally adapted seed. The means to achieve this have been discussed and center on linkages between the breeding program and the farmer through the extension service. Once superior varieties with good characteristics for stability and level of yield are available, the matter of raising demand becomes a problem for seed policymakers rather than breeders. Lack of demand for improved seed may be attributable to economic factors or to cultural and institutional factors. When land tenure, infrastructure, market access, or credit availability are limiting factors, then these must be attacked directly if demand for seed is to expand. Even if superior seed is available and farmers would like to use it, effective demand will not materialize unless farmers can sell their crop at a profit and improve their standard of living.

The ultimate obstacle to increasing demand is the farmer. As described in Chapter 6, small farmers are reluctant to depart from traditional, proved ways. Traditional farmers are distrustful of changes because they know many things can go wrong. As a consequence, the seed industry has a grave responsibility to conduct realistic trials to ensure that the new varieties are truly superior over a broad range of moisture and input levels. Only when this has been ascertained should new varieties be multiplied up for commercial sale and introduced to farmers in demonstration plots. The extension service should promote the new varieties, but be realistic in explaining what other inputs are needed and what yields are likely to be. A large part of the burden of overcoming farmer reluctance to use new seed falls on the extension service. The strengthening of this system must have a high priority if agricultural growth and seed industry development are to show adequate progress among small farmers.

One of the most common methods of creating new demand for improved seed is to subsidize its production and to sell it at less than full cost. This approach does raise demand and helps move a system from Stage 1 to Stage 2. Unfortunately, in the process of raising demand, production subsidies make matters more difficult in other areas and slow down progress from Stage 2 to Stage 3. The first problem with all subsidies is that they must be financed by tax revenues, public borrowing or credit creation, and all these paths are increasingly unacceptable to financially strapped governments. In practice, production subsidies have usually led to an overly large, inefficient public seed sector. When a significant share of funds for the seed program is subsidies, the incentives to ascertain and serve the farmers' requirements are weakened. Decisions tend to be centralized, based on what will satisfy the government and justify continued subsidies rather than on what will satisfy farmers and raise sales. The weakened incentives to respond to market forces result in slower development of important seed operations, such as quality control, distribution, promotion, and extension. Public sector targets are usually set in terms of bulk amounts of seed produced, and in meeting those targets and justifying the subsidy, quality slips, distribution and promotion become afterthoughts, and the extension system is not emphasized as it should be.

Subsidies are also innately hostile to private sector development. They depress prices, making it very difficult for a private organization to cover even direct costs, let alone generate a return on capital. In such a situation, tax incentives have no meaning because there are no profits to be taxed. There may also be a tendency for a subsidized public sector to try to expand its activities vis à vis the private sector. Where the private sector is weak or in decline, it could be argued that the public sector needs to expand to assure an adequate supply of improved seed; such expansion requires a larger subsidy and more people to do the job. Action along such lines will further weaken private seed organizations and ultimately end their activities. The result would be a loss of competition between seed producing organizations, dependence on the public sector for the full range of seed operations, and, as experience shows, the likelihood of a high cost seed industry that does not meet the farmers' needs. Thus, production subsidies at the early stages of seed industry development tend to create conditions that may accelerate progress in the short run but make it more difficult to sustain progress in the middle stages.

There is no disagreement on the matter of farmers' reluctance to pay market prices for improved seed that they feel are inferior to their saved seed. However, production subsidies for the public sector, with all the accompanying problems, are probably the worst way to reduce costs to farmers. One improvement would be to make subsidies available to private and public seed producers on the basis of seed units actually sold. If the farmers do not buy the seed, the producer does not receive the subsidy. This would reduce discrimination against private producers and, more important, force the seed industry to concentrate adequate attention on quality control, distribution, promotion, and follow-up services.

Another approach that could be implemented where rural credit institutions exist is to use favorable credit terms for subsidizing seed purchases, as in the south of Brazil. Seed is priced and sold on a full cost basis, but farmers are subsidized through attractive credit terms for their purchase of improved public or private seed. This system would reduce any potential for seed organizations claiming seed sales in excess of what was actually achieved so as to get higher subsidy payments. Perhaps the best approach is to subsidize the production of improved seed through public research and development and release the basic seed to responsible public and private organizations and growers. This strategy may result in slower initial rates of growth but, as shown in Kenya, it lays down a better foundation for more rapid progress on the supply side during later stages of seed industry development.

Whereas raising demand is the key to moving from Stage 1 to Stage 2, progress during later stages is characterized by successful supply side adjustments in terms of the quantity, quality and timeliness of seed supplies. The production of a large quantity of "improved" seed is the easiest part of tasks facing seed suppliers; the difficult challenge is to get the correct mix of seed required, to maintain seed quality through the entire chain of operations, to deliver the seed to the farmer on time, to get the farmer to pay a full cost price, and to enable the farmer to show a profit. Unless all these conditions are met, the farmer will not return to purchase new seed, the seed industry will not continue to develop, and agricultural progress will be unsatisfactory. The proper provisioning of improved seed is a difficult, complex task requiring extensive teamwork. The breakdown of one operation can undermine the efforts of everyone throughout the chain and, for this reason, the incentive structure must clearly be focused on satisfying the needs of the farmer. The greatest weakness of the public sector is the lack of appropriate and immediate incentives within the bureaucratic system, and experience shows that excessive reliance on public seed producers is probably the most important obstacle to progress in the middle stages of seed development.

The discussion of programs and policies recommended to build up the supply capacity of the seed industry will be brief,[31] focusing on four areas:
- multiplication and growing seed for commercial sale;
- seed conditioning, processing and storage;
- the distribution and marketing of seed;
- informational, promotional and follow-up services by the seed producer.

These areas seem to present the most difficulties for seed policymakers in developing countries. They are also the keys to getting improved seed supplies out to the farmer and into agricultural production.

One of the first supply bottlenecks encountered in seed production is having enough farmers who are willing and able to multiply up foundation seed. This can be a problem in developed countries, and among the developing country cases, India and Brazil (in the northeast) certainly had difficulties in carrying out this operation. Risk and compensation are very important in having a good group of contract growers. In India the growers seemed to be taking a disproportionate share of overall risk and their return was inadequate. Part of the risk was related to weather, as in Brazil, but part of it was attributable to vagaries in the process of accepting the seed that had been multiplied up. In a proper seed program, growers need to be recruited, trained and adequately compensated if the multiplication operation is to increase production and maintain quality. Over time, there must also be the capability to adjust the mix of seed grown to meet changing needs for finished seed. The training of growers in the techniques of multiplication, including roguing and harvesting, should be a government responsibility during initial stages of development. Further, for a fee that covers costs, growers should be assisted with services providing input applications or management. Finally, to retain experienced seed growers, compensation and costs should be structured to generate a fair return to the growers' land and labor. As the seed industry expands, there will be an ongoing need for more contract seed growers, and the recruitment, training and assisting of these people should be an integral part of the overall national seed strategy.

Physical facilities for drying, sorting, conditioning, packaging, and storing seed should be expanded over time as part of a planned growth strategy. In many developing countries such facilities are inadequate, relative even to current needs, and quickly become critical problems if not provided for. They should be located to minimize transport costs and to allow maximum flexibility in servicing various areas as demand levels grow and demand patterns change over the course of time. Such decisions are quite difficult because it is almost impossible to forecast the quantity and mix of seed that will be demanded in three to five years in various regions of the country. Nevertheless, decisions on the construction and location of seed facilities are extremely important and require careful planning. Whenever possible, it is desirable to have the private sector handle these operations, make the tough decisions and face the consequences in terms of success or failure. Support for seed industry expansion in the form of favorable credits may be a means of encouraging the private sector, but the costs and risks should be borne by the private investor. It goes without saying that private sector expansion will not occur unless there is the strong likelihood of profits over time. Where the private sector is unable or unwilling to step in and furnish the facilities required, then public sector action should be very carefully planned and executed.

The distribution and marketing of seed is perhaps the most difficult and time-consuming operation in the whole process. As noted in the discussion of the Kenya Seed Company, establishing a satisfactory distribution network took much longer than anticipated. In addition, because distribution takes so long to get right, it is a mistake to adopt policies at an early stage that will inhibit or defer its development. The first aspects of distribution are packaging and storage. The seed units must be in sizes appropriate for the market, labeled clearly so that the product is easily recognized, and well sealed. The sealing process ensures that the contents have not been tampered with and that deterioration has been minimal during storage. Storage facilities are important and, particularly in the tropics and subtropics, humidity and temperature must be kept down to acceptable levels. With regard to actual distribution, there is no single formula that works in all cases. In Kenya, seed reached small farmers through existing commercial distribution networks on a cash basis. In Brazil, seed purchases were provided with credit subsidies and the extension system was heavily involved. In other cases, distribution through networks of farmer-dealers have proved to be extremely successful. There does not seem to be any substitute for trial and error and learning quickly from past mistakes.

It is clear that the distributional system must meet the farmers' needs in terms of the appropriate seed delivered at the appropriate time. The right location is important, but perhaps less critical; Thai farmers are willing to travel some distance to purchase seed. In addition, the distribution system must perceive the changing needs of the farmers and pass this important information back up the chain of seed industry operations to policymakers. Finally, and perhaps most important, the product must be sold at a price that covers all costs and provides enough profit for distributors to encourage them to promote the product among local farmers. The final distributor is a critical part of the seed industry; unless the distributor can convince the farmer to make the purchase, the effect of improved seed will not be reflected in greater agricultural production.

There is also an informational and promotional function that needs to originate with the seed producing organization. Information on the timing and characteristics of newly available varieties should be provided to the seed distributors and the extension system. Demonstration plots should be set up for viewing by extension officials and interested farmers. Follow-up investigations should be conducted by seed producers to see how farmers are doing with their products. This will help to identify the range of conditions under which the variety performs well or performs poorly. These investigations can also help to identify new products that farmers need. Such information is invaluable to a conscientious seed producing organization, but an effort has to be made to seek it out. When the organization's success is determined by sales to farmers, such efforts are more likely to be made.

Quality control and the regulation of the seed production process is typically one of the weakest parts of the seed programs in developing countries. Even when laws exist, implementation and execution are inadequate. Facilities, staffing and training are probably unequal to the task at hand in the majority of developing countries. Another difficulty is the existence of a special relationship, such that public sector producers get more expeditious and perhaps favorable treatment than the private sector. The objective should be an arm's length relationship between the certifying and regulating agencies, on the one hand, and the public sector seed producers on the other. Staffing, equipment and facilities should be adequate to carry out their responsibilities quickly and correctly. The worst possible situation involves extreme regulation of seed producers' activities with very little positive effect on the quality of the seed produced. This is a prescription for high costs, the destruction of the private sector, and weak seed quality control. The best possible situation involves adequate seed quality and a minimum of regulation. Brazil and Kenya may be the best examples of the latter case. Seed quality standards should not be set at unrealistically high levels, particularly during the initial stages; they should be set at adequate levels and raised slowly as the seed industry becomes more advanced. Beyond certification, other regulations should be kept to an absolute minimum and enforced even-handedly at all times between public and private breeders and producers.

In every country the situation is somewhat different and calls for a different strategy for seed industry development. Therefore, no single recipe for success can be provided for policymakers. However, there are a few general principles that seem to be widely applicable:

(1) work from a clear and consistent strategy that takes at least a medium-term perspective on the seed development process;

(2) create a policy environment and incentive system in which the private sector does as much of the investing and risk taking as possible;

(3) focus public resources on operations, crops and regions that do not promise sufficient return to attract private activity;

(4) training programs are a particularly good use of public sector resources;

(5) keep seed regulations and subsidies to an absolute minimum and utilize market forces whenever possible.

These principles do not argue for a laissez-faire attitude toward seed industry development. On the contrary, the government has to be involved in designing and implementing an appropriate national strategy.

The principles argue for an awareness of the strengths and weaknesses of public and private seed organizations, as revealed by experience. And once aware of these, policymakers should aim to enhance and build on strengths and work around weaknesses. Policymakers must be realistic about what the public and private sectors can successfully accomplish and plan accordingly.

FOOTNOTES, CHAPTER 7

1. Satya Dera, "The National Seed Project in India," *Journal of Overseas Development* (October 1980), Vol. XIX, No. 4, p. 263.
2. World Bank, *World Development Report* (WDR), 1982, p. 76.
3. Ibid.
4. Dera, "The National Seed Project," p. 263.
5. Ibid., p. 264.
6. Ibid., p. 268.
7. Ibid.
8. WDR, 1982, p. 76.
9. Ibid.
10. For a more detailed discussion of many of these issues, see Lois A. Gram, "Investment in Thailand's Seed Sector, A Public-Private Continuum" (World Bank, mimeo, 1983).
11. Maxwell L. Brown, "Some World Bank Experience in Financing Seed Projects with Special Reference to Asia," paper presented at the CIAT Conference on Improved Seed for the Small Farmer, Cali, Colombia, August 1982.
12. For more details on KSC operations, see *Report on the Kenya Seed Company*, E.J.R., AGP/SIDP/81/61 (FAO, March 1981).
13. W.H. Verburgt, "Increasing Seed Sales, Especially to Farmers with Small Holdings," p. 1, paper submitted to the CIAT Conference on Improved Seed for the Small Farmer, Cali, Colombia, August 1982.
14. Ibid., p. 3.
15. Ibid., p. 1.
16. J. Sneep and A.J.T. Hendriksen, *Plant Breeding Perspectives* (Wageningen, Netherlands: Center for Agricultural Publishing and Documentation, 1979), p. 393.
17. Ibid.
18. Edwin M. Kania, Jr., *Pioneer Overseas Corporation Case Study* (Cambridge: Harvard Business School, Case 4-583-070, 1982), p. 20.
19. See Lois A. Gram, "The Evolution of the Turkish Seeds Industry" (World Bank, mimeo, February 1984).
20. Ibid., p. 12.
21. Figures from various publications, Direcion General de Economia Agricola, Secretaria de Agricultura y Recursos Hidraulicos.
22. For more detail on the problems, prospects and policies of Mexican agriculture, see the *Programa de Alimentacion, 1983-1988* (Mexico City: Government of Mexico, October 1983).
23. Various publications of PRONASE, national seed producer of Mexico.
24. Ibid.

25. Data in this section from Brazilian Ministry of Agriculture.
26. U.S. Department of Agriculture, Foreign Agricultural Service, "Brazilian Agricultural Situation" (February 1984), mimeo, AGR Number BR 4603.
27. Falvio Popingis, "Genetic and Basic Seed Situation in Brazil," CIAT Advanced Training Program, Cali, Colombia, 1980, mimeo, p. 1.
28. Program for Increasing Production and Utilization of Improved Seeds 1980/84.
29. Popingis, "Genetic and Basic Seed," p. 31.
30. This section is based on a paper prepared by Lois A. Gram, "ICTA: Entrepreneurship in the Guatemalan Seed Industry" (World Bank, mimeo, 1984).
31. For more on the techniques of seed production and management, see Johnson Douglas' book, *Successful Seed Programs, A Planning and Management Guide* (Boulder, Colo.: Westview Press, 1980).

Barriers to International Trade in Seed | 8

INTRODUCTION

The focus of this chapter is on tariffs and nontariff barriers to international trade in improved seed. NTBs are complex and difficult to evaluate. In principle it is possible to calculate the quantitative effect of NTBs on trade patterns, but for seed the required data on demand and supply conditions within specific markets are not available. The objective here, therefore, is necessarily limited to providing an understanding of:

    (1) two qualifications to normal free trade principles that are relevant in making seed trade policy;

    (2) the kinds of barriers that countries utilize to inhibit trade in seed;

    (3) the consequences of a system of barriers that prohibits or severely reduces trade in improved seed.

Little or no analysis of barriers to international trade in seed has hitherto been done. Prior to the Tokyo Round negotiations, barriers to trade in agricultural products were not documented systematically. The current system is based primarily on a program of "self reporting," whereby countries list the barriers to agricultural trade that they employ; this is supplemented by a listing of complaints by exporters. No one familiar with agricultural trade views this system as comprehensive. For seeds, the problem is made worse because of the critical importance of technical standards that are widely felt to be chronically underreported in the General Agreement on Tariffs and Trade. Therefore, this analysis will serve primarily to illustrate and illuminate the existence and consequences of tariffs and nontariff barriers to seed trade.

Barriers to seed trade differ dramatically between imports of germplasm (seed for use in breeding programs) and imports of commercial seed for sale to farmers. Germplasm is almost universally allowed free access into countries. Commercial seed, in contrast, usually faces significant trade barriers, especially among developing countries. This discussion thus pertains to commercial seed imports, unless otherwise specified.

There are both natural and manmade barriers to trade. Natural barriers typically include the transportation and shipping costs of engaging in international trade, which add to the cost of a good exported rather than retained for domestic consumption. Another natural barrier to trade of particular importance to seed is the adaptability of certain varieties for use in different agro-environments. As described in Chapter 2, many varieties, especially cereals and forage crops, cannot be utilized in latitudes different from the region for which they were bred. Other varieties may be more adaptable, but are still at risk regarding resistance to pests and diseases, moisture conditions and other local environmental factors. Only with vegetable seed can it be presumed that local adaptability may not be a problem. For most other crops, the adaptability factor clearly limits the potential export market to specific regions. American corn varieties will perform well in Italy, Spain and southern France, but not in northern France or Germany, which are in colder latitudes. Because transportation costs only raise the price at which seed is available and are low relative to their value added, the adaptability factor is probably the biggest natural barrier to wider international trade in improved seed.

Manmade barriers also affect seed availability and/or prices. Tariffs and other barriers that raise the price of imported seed are usually not high enough to reduce trade significantly below what would have otherwise occurred. When reducing imports is the policy objective, most countries opt for nontariff barriers that directly limit or preclude availability rather than rely on tariffs and the price mechanism. There are important differences among types of economies regarding their use of trade barriers. The advanced market economies, except for Australia, do not have prohibitive trade barriers, and as a result seed trade flows among these countries are well developed. Among the centrally planned economies, international transactions are monopolized by state trading agencies and vary greatly from year to year. The developing countries use a wide range of NTBs to control trade in seed; total prohibition of most types of seed imports exists in many of these countries. In addition, trade barriers tend to have a negative effect on the quality of seed available to the farmer. As discussed in Chapter 7, prices in the developing countries are generally held down through domestic subsidies or other devices, yet seed availability and quality are often below par. Fewer trade barriers would help with these problems. Increased freedom of trade is an important aspect of a successful strategy for seed industry development, especially during the later stages, and it can ease many of the seed problems facing both developing and developed countries. It increases availability of new varieties, helps to improve quality, provides competition for domestic producers, and broadens a country's access to the world's pool of germplasm and research.

In an idealized, perfectly competitive economic system, trade barriers, domestic market distortions and externalities are absent, the consumer can choose freely between foreign and domestic products, the prices of identical products are equal, prices represent true resource costs, and producers' costs equal society's costs. The introduction of trade barriers upsets these relationships and creates inefficiencies. In particular, trade barriers tend to raise prices and reduce the quality and availability of products to the consumer. The consumer and the society end up paying more than the true resource cost of the product. Some of the higher cost paid by the consumer is redistributed to domestic producers and some is lost in terms of waste and misallocation of resources. Trade barriers, in their net effect, tend to redistribute income away from consumers toward domestic producers of the protected products and to reduce the aggregate income of the whole society. These results are most assured when domestic market imperfections, including unemployment and externalities (i.e., side effects not accounted for in market prices and costs), are at a minimum.

The seed industry is one of a few areas in which externalities are significant relative to total economic activity. Two aspects of seed production create important qualifications to customary free trade principles:

> (1) a large part of the value added in improved seed is knowledge that can be embodied in the seed at very low cost once the initial research expenses have been met;
> (2) open-pollinated seed can be reproduced by the farmer at little additional cost after the improved seed has been purchased once.

In a situation without market distortions and externalities, there should be an equilibrium where, at the margin, the cost of producing the good equals the price that also equals the value of the benefit received by the purchaser (marginal cost = price = marginal benefit). For improved seed these equilibrium relationships do not generally hold because of the two externalities just noted. At the margin, the cost of producing improved seed is very low because once the genetic knowledge has been developed, the marginal cost of producing additional improved seed does not include the fixed research cost. Moreover, the ease with which farmers can reproduce improved seed by purchasing the new variety once and saving some of their crop for sowing the following year — the self-production externality — ensures that, for open-pollinated crops, the improved seed producer cannot charge much more than the marginal cost of seed production. Plant variety protection will raise the price because it reduces "free riders" to farmers producing seed for personal use, and it ensures that other seed producers cannot directly exploit and sell the variety resulting from the original research. However, PVP certainly does not eliminate the problem; there is no way to prevent farmers from satisfying their future needs at very low cost after the initial purchase of improved seed. This situation consequently ensures that marginal benefit will almost always exceed marginal cost and that price will tend toward marginal cost for open-pollinated seed. The knowledge externality causes the divergence between the marginal cost of producing seed and the marginal benefit.

The knowledge externality pertains to hybrid seed as well, but the self-production externality does not. The price is, therefore, going to be higher than for open-pollinated seed. Because the hybrid seed cannot be easily replicated by farmers or anyone else, the producer is able to keep prices well above the marginal cost of production and to cover the fixed costs of conducting research and breeding new varieties. Even with hybrid varieties, however, evidence indicates that the price of seed is nowhere near the level of the marginal benefit to the farmer. A wide range of studies by hybrid seed producers indicate that farmers do not move strongly to accept new varieties until the benefit reaches a range of two and one-half to four times the price of the seed. For open-pollinated seed, the range of benefit ratios resulting in acceptance is even higher.

As a result of these two externalities, private seed producers, especially of open-pollinated varieties, have a chronic inability to pay for genetic research at levels that are desirable from the point of view of the consumers of agricultural products. This means that, left solely to private seed producers, genetic research would be significantly underfunded. Therefore, there is a strong argument based on economic efficiency for subsidies to plant breeding research. Two common forms of subsidy are publicly supported research by public breeders and programs funded by foundations or international institutes. Ordinarily, subsidies have the effect of distorting markets, creating trade barriers and causing inefficiencies. In the case of seed, especially open-pollinated seed, the presence of the externalities already discussed creates a quite different situation in which subsidized research does not lead to a misallocation of resources and is highly desirable. The problem arises in releasing the research results. Strictly speaking, the products of research funded by public subsidies should be given away to all—private companies as well as public agencies—to compensate for the knowledge externality. A more limited release of new varieties produced through publicly subsidized research is undesirable and should be reviewed. Specifically, a policy of restricting production and distribution of subsidized varieties to public seed producers constitutes a serious distortion of seed markets and a clear barrier to trade.

TRADE BARRIERS

Traditionally, agriculture has been impeded in most countries by extensive trade barriers. Attempts to reduce them have met with little success, although they have been subject to continuous scrutiny and discussion, both within the multilateral GATT framework and bilaterally. By comparison, industrial products have experienced almost continual tariff declines over the past four decades, so that tariffs are no longer significant barriers to trade in those products among the developed countries. Nontariff barriers have become relatively more important for industrial goods, but for the United States at least, even they seem to have declined in the early 1980s.[1] In contrast, agricultural barriers to trade remain formidable, particularly the nontariff barriers, with agricultural subsidies for domestic production and/or consumption being the most prevalent.

International trade in improved seed is likewise characterized by considerable nontariff barriers obstructing the workings of the market. In addition to domestic subsidies, phytosanitary (health and cleanliness) standards are highly prevalent barriers to international seed flows. While necessary as a means of protecting against the introduction of diseases and pests that could seriously damage domestic crops, these standards are frequently used to prevent or obstruct trade when there is no genuine phytosanitary problem. A sample of about 20 developed and developing countries has identified more than two dozen practices that could be construed as barriers to, or distortions of, trade. This list is presented as Table 8-1 and is intended to be as comprehensive as possible, although it is certainly not all-inclusive. The various nontariff barriers and distortions are organized into five categories: import controls, health and performance standards, structural and economic barriers, political barriers, and export subsidies.

TABLE 8-1. LIST OF BARRIERS TO INTERNATIONAL TRADE IN SEED

Tariffs
Nontariff Barriers
   A. Import Controls
      1. Variable levies and special charges
      2. Quotas and prohibitions
      3. Import licenses
      4. Domestic content restrictions
      5. Domestic processing requirements
      6. Production/import ratios
      7. Actions based on various GATT articles

   B. Health and Performance Standards
      8. Phytosanitary regulations
      9. Certification
     10. Tests, proof of superiority

Continued

TABLE 8-1 Continued

   11. Cataloguing, inscription
   12. Packaging and labeling restrictions
   13. Documentation requirements

C. Structural and Economic Barriers
   14. State trading agencies
   15. Inconvertible currencies
   16. Barter trade
   17. Price controls
   18. Marketing and distributional restrictions
   19. Domestic research subsidies
   20. Domestic production subsidies
   21. Domestic credit preferences

D. Political Barriers
   22. Boycott
   23. Embargo on exports of germplasm

E. Export Subsidies
   24. Tied aid and grants
   25. Gifts in kind
   26. Foreign exchange or credit preference for exports
   27. Export production subsidies
   28. Dumping
   29. Export research subsidies

In spite of this imposing list, international sales of seed have expanded dramatically (see discussion in Chapter 4), testifying to both the persistence of seed suppliers and the strength of demand for improved seed. There is some evidence of seed entering countries illegally where barriers to trade were enacted but could not be enforced. A consideration of these barriers to trade, beginning with a brief discussion of tariffs and proceeding to the five categories of nontariff barriers, follows.

Tariffs

Tariffs are a tax imposed on imported goods as they enter a customs union; this is usually a single country or may be a "common market" made up of several countries with common external tariffs and no tariff among the member countries. The tariff may be a fixed amount per unit of imported good, e.g., $10 per 100 pound bag of soybean seed, or may be ad valorem, i.e., a percent charge on the value of the imported item. More common than flat charges, ad valorem tariffs rise in value automatically and thereby raise tariff revenues as inflation raises the price of imported goods. However, ad valorem tariffs are based on a declaration of the value of the goods, and this can result in disputes between the importer and the customs agency charged with collecting tariffs. Examples of tariffs for hybrid corn seed and seed of other cereals are presented in Table 8-2. In some cases countries do not separate out seed imports in their tariff schedules, so the data shown may be for "other" imports of corn or cereals. Only the United States denominates its tariff charges on a fixed unit basis; all the others use ad valorem rates.

For many countries, tariff levels are low or zero, but for others the rates are moderately high and pose real barriers to trade, as in Spain and several Latin American countries. In Spain, "superior" seed (except for hybrid corn) has zero tariffs, while hybrid corn and "other" seed face very high rates. It was noted in Chapter 5 that the EC's 4 percent duty on corn seed was not of consequence to trade flows, but the higher rates imposed by Western Europe on other types of cereal seed are a matter of concern. The tariffs employed by Japan seem rather high, more in line with Latin America than a developed country. Tariff rates on industrial goods for developed countries tend to cluster in the range of 2.5 to 8 percent, which is not far off the tariff

TABLE 8-2. EXAMPLES OF TARIFFS ON
SEED IMPORTS, 1980-82

	Hybrid Corn	Other Cereals
Argentina	12%	12%
Australia	0	0
Austria	0	0
Colombia	15%	15-20%
EC	4%	13-20%
Egypt	0	0
Japan	10%	10-15%
Mexico	0	0
Peru	5%	5%
Spain	30%	0-45%[a]
Sweden	0	0-2.5%
Uruguay	10%	10%
U.S.	3¢[b]	0-5%
Venezuela	20%	10-20%

a. 0 for seed of "superior quality."
b. 3¢ per bushel of 56 pounds.

Source: Country tariff schedules as provided by the GATT.

rates for seed imports into the same countries. Developing countries have typically higher tariff structures on both industrial goods and seed imports. Most of the developing countries listed in Table 8-2 have tariffs of 10 percent or more, which is less in general than their tariff structure for imports of industrial goods. On balance, however, tariffs on seed are not prohibitive except in a few instances—for example, hybrid corn imports to Spain. Tariff duties of 10 to 15 percent, while usually not enough to keep out imports, do reduce the level of trade and competition in domestic markets.

Countries with high tariffs tend to have even more powerful non-tariff barriers against seed imports, and countries with low tariffs have low nontariff barriers. In both cases the nontariff barriers are the more significant.

Import Controls

The first type of NTBs to be considered aim directly to monitor or control the inflow of imports through administrative devices, although the rationale may be broader, e.g., to budget foreign exchange or to be able to forecast market balance. Also in this category are requirements that force an importer to produce or process domestically to be able to import.

Variable levies and special charges are similar to tariffs but can be styled to impact selectively on certain types or classes of imports. One example is the reference price mechanism used by the EC to impose variable levies on hybrid corn seed imports, as mentioned in Chapter 5. Each year a reference price is calculated based on historical price patterns. If imported hybrid corn seed is valued at less than the reference price, which is a minimum import price, a levy is imposed to bring the price up to the reference level. Thus, the lowest cost producers of corn seed pay the highest levies and the mechanism acts to remove, or at least severely limit, price competition from importers. Table 5-4, p. 155, shows the range of price differentials that prevail and the impact that variable levies could have, particularly on Eastern European producers. In fact, there is a GATT limitation of 4 percent on the import levy, so that GATT members are not subject to the full effect of the levy. Non-GATT exporters of hybrid corn seed to the EC must pay the full levy when their seed is priced lower than the calculated reference price.

Reference price determinations are among the most common types of variable levies, and there are many variations on this theme of nontariff import taxes. A related problem occurs when customs evaluates an import at the highest possible tariff rate when any discretion is involved. Sorghum, for example, can be used as a grain or a forage (sweet sorghum), but customs officials tend to impose duties on specific shipments of sorghum seed imports at the higher rate, irrespective of whether the seed is for grain or forage use. These are just two of the many ways in which charges and levies are calculated and imposed on imports.

Quotas and prohibitions are straightforward legal limitations on the amount and type of goods that can be imported. The motivation for imposing them is usually a combination of a perceived "shortage" of foreign exchange to pay for imports at the current exchange rate; a desire to encourage domestic production of the goods; and a bureaucratic desire for control and predictability in managing domestic and foreign economic affairs.

Tariffs are unpredictable tools in managing an economy and particularly a country's foreign exchange budget. Even when they are high, demand for imports may draw in an undesirably high level of foreign goods, creating shortages of foreign exchange and international financial problems for the country's policymakers. Quotas are a more certain and more precise means of hitting targets. Therefore, among developing countries with foreign exchange and international debt problems, quotas are prevalent and have extended to international trade in seed. It is probably accurate to say, regarding seed, that true import-export markets do not exist in the developing countries, most international trade is administered, and the public sector is usually the dominant importer. Quotas in one form or another are thus the rule, rather than the exception, for international seed trade with developing countries.

Import licenses are a common means of imposing quotas, but they are usually introduced under other guises. Import licenses for hybrid corn were introduced in the EC in the late 1970s. Ostensibly they were a means of forecasting import levels and thereby estimating the overall balance of supply and demand, i.e., improving "market transparency." Complaints about increases in imports were emerging at that time, and it appears that the import licenses were also viewed as a "compromise to show some type of support for the European industry without restricting trade."[2] The basic elements of the EC program for hybrid corn include:

- automatic and rapid issuance of import licenses upon request by the importing country;
- six-month validity for seed imported as part of a seed multiplication contract;
- three-month validity for regular imports plus a small deposit based on the volume of imports, to be refunded when the license is complied with.

All the hybrid corn imports under contract are by EC-based companies, while almost all the rest are by companies from other countries. Therefore, in terms of time and the required deposit, the EC importers are effectively favored.

Import licenses in Latin America are the key administrative tool used to limit all seed imports and impose effective quotas, so their introduction by the EC was viewed with misgiving. EC officials assured foreign seed officials and producers that the import license scheme was not intended to restrict trade, but to increase generally available information on expected import levels. Some European seed producers were privately skeptical, but experience through 1985 has shown that the licenses have not been used to restrict imports. In some instances licenses were held up, but it is not clear whether this was intended by the officials of the country or due to the uneven pace of the issuing process. In the case of hybrid corn seed imports into the EC, the import licensing system is viewed as an annoyance, but to date it has not proved to be a real barrier to trade. In other settings, particularly in the Third World, import licensing systems are an integral part of a series of strong barriers to free international trade in improved seed.

Domestic content and processing restrictions and domestic production/import ratios are alternative means of accomplishing the same goal. Imports are not allowed unless domestic economic activity is at a certain minimum level. Trade barriers of this kind can be found in many types of manufacturing and agricultural areas. Domestic content legislation has, for example, been introduced in the United States to force auto importers to move part of their operations to the United States. In Spain and Austria, corn seed imports are restricted to a fixed ratio of domestic production, one-fourth in Spain and two-thirds in Austria. In both countries, the objective is to force seed producers to set up local operations and develop local sources of supply regardless of the cost disadvantages. These practices seriously distort trade, and countries should be strongly discouraged from using them. If applied literally, they would severely harm agriculture if domestic seed production were to fall off because of poor weather or disease. Such restrictions would most likely be relaxed if a need for imported seed were perceived, but under the best of circumstances they eliminate smaller seed importers, restrict trade in seed and reduce competition.

As was seen in Chapter 5, if the proper conditions apply, countries may take action under the GATT to establish trade barriers. Actions invoking GATT's Article XIX, the Safeguard Clause, must, for example, meet the criteria of unforeseen developments resulting from compliance with GATT obligations due to increased imports that threaten seriously to injure domestic producers of competitive products. The presumption is that the action will be temporary (that the industry can recover and be competitive in time), and that compensation in the form of trade concessions in other areas can be required by the exporting countries. These conditions are rather specific and provide fairly tight guidelines for the invocation of an acceptable Article XIX action to erect a barrier against imports. Such barriers, albeit temporary and limited in scope, are part of the price of obtaining the benefits of the relatively free trade that results from the GATT system. Nonetheless, import controls put in place as a result of a safeguard action must be included among the types of trade barriers that could restrict seed imports.

Health and Performance Standards

Phytosanitary restrictions are common the world over and are an essential part of a country's defense against the introduction of diseases and pests that could devastate domestic crops and other flora. Tourists are familiar with the forms concerning plants and fruits that must be filled out when traveling overseas. Even within countries, there may be agricultural inspection stations at the borders of a protected state or region, as in traveling to California in the United States. The need to maintain restrictions on the introduction of potentially dangerous plant material and to inspect seeds and plants being transported internationally is universally recognized. However, while phytosanitary regulations are essential, care must be exercised in their legislation and enforcement so that they are not unnecessarily restrictive and become a barrier to imports.

Phytosanitary regulations are complex and technical, but a detailed discussion of several examples will show how such regulations can become trade barriers. Seed producers and officials who deal with international trade in seed probably spend more time and effort coping with the legal and administrative problems arising from phytosanitary, certification, cataloguing, inscription, and other health and performance standards than from all the other categories of barriers. In addition, these regulations are constantly evolving; countries may feel they should require new types of inspection, reword the language of the existing permit, change packaging or labeling requirements, or prohibit imports from a country felt to contain a newly identified plant disease. Seed producers and exporters and their representatives have to be constantly aware of these changes or face significant losses when their seed shipments are rejected. This is an ongoing problem that is at best an obstacle to trade and at worst an insurmountable barrier.

One example of a phytosanitary regulation having the effect of a trade barrier was the 1979 EC directive restricting the importation of alfalfa seed. England and Ireland long had restrictions on imported alfalfa seed because of their desire to avoid a type of bacterial wilt that affects alfalfa. The problem was thought to be present but not widespread in England. To stamp it out, strong restrictions on alfalfa seed were enforced. When England joined the EC in the 1970s, some English phytosanitary restrictions were adopted by the whole of the EC and a directive was issued on bacterial wilt. To be acceptable for importation, alfalfa seed had to pass four tests and be officially documented as having passed. Two problems resulted: the EC tests were more restrictive than England's had been, and no agency existed in the United States or elsewhere that was prepared to undertake the required investigations to document that the seed and field were free of bacterial wilt.

The effect was to exclude a large share of U.S. alfalfa land from consideration for growing seed for the EC market and to reduce dramatically U.S. exports to Europe. Ironically, bacterial wilt already existed in several European countries, including most of Eastern Europe, Germany, Italy, and probably France. Moreover, England was no longer trying to eradicate the wilt, but to develop resistant alfalfa varieties.[3] Thus, because the disease was already present, the net effect was almost entirely protectionist, and the restriction served only as a barrier to trade. The problem has dragged on since 1979, and no resolution is in sight. There is no legitimate scientific basis for the phytosanitary exclusion, yet the directive remains in effect because of bureaucratic inertia. The most recent development allows varieties that are resistant to bacterial wilt, but no one is sure of how to guarantee such resistance. It will be at least several years before tests are completed and resistant varieties identified, documented and accepted by the EC cataloguing system. Although most EC seed industry officials privately recognize the problem, once regulations are in place it is very hard to get them removed. Thus, a spurious phytosanitary regulation has created 5 to 10 years of a severe trade barrier.

An example of a phytosanitary restriction combined with a language problem is the case of tomato seed exports to Italy. The EC phytosanitary regulation on tomato seed requires that the seed come from regions known to be free from a particular virus, or that no symptoms of the virus have been observed since the last crop. A shipment of tomato seed sent to Italy was held up because, although it came from a region (California) recognized as virus-free, no field inspection had been made to certify the absence of symptoms during the last crop year. In the Italian translation of the EC directive, the word "or" was omitted and therefore both conditions had to be met. Under EC rules, one member government can be more stringent regarding third country imports but must follow the EC rules for trade among the nine member countries; thus, tomato seed imported to Italy from outside the EC must meet more stringent tests than imports from other EC countries. As a result, the tomato seed shipment was detained at the Milan airport while the discussion on directives and language proceeded, although the shipment was finally released by the Italian authorities. In the future, tomato seed imports from outside the EC must meet both tests or be denied entry to the Italian market.

The United States has extensive phytosanitary restrictions on the importation of seed and other plant and agricultural products, some of which are unnecessary. Soybean seed imports, for example, must be sprayed to kill spores that carry soybean rust disease, but it is unlikely that the spores are even present on the seed. Thus, the spraying requirement means extra expense and difficulty for the exporter with little or no genuine phytosanitary benefit. Another example of phytosanitary overkill is the U.S. restriction on alfalfa seed imports from Europe due to fungal wilt. This is almost a perfect parallel to bacterial wilt because it appears that fungal wilt now exists within the United States, and there is no longer the need to control alfalfa seed imports that could transmit this disease. The similarity of the trade barriers suggests a quid pro quo whereby both the EC and the United States do away with their unnecessary restrictions and thereby benefit from the greater choice of varieties and improved competition in Europe and the United States.

Phytosanitary regulations are critically important to the protection of a country's ecology and agricultural endowment. Such regulations are thus extremely widespread and have enormous potential to create undesirable barriers to trade. There is a fine line, however, between the scientifically justified safeguarding of a country's plant resources and the economically unjustified creation of barriers to trade. Such barriers in the name of phytosanitary controls are extensive, and bureaucracies have been too slow to amend the directives that have been shown to be unjustified.

Closely akin to phytosanitary regulations are the various seed production control measures that producers must take and document if their seed is to be certified. Certification requirements vary considerably among countries, but the types of controls generally utilized include: field inspections for varietal purity, the absence of diseases and the distance from fields of other varieties (to prevent undesired cross-pollination); seed testing to verify identity and determine other key characteristics such as moisture content, freedom from weed seed and inert materials, and germination rates; and seed treatment with appropriate chemicals for protection against diseases and insects. When these procedures have been carried out and verified, the seed is certified as having the approval of the national seed certifying agency.

For domestic producers, the seed certification process is usually well known, and inspectors are at hand to check that the proper seed production controls are in place. These programs can be an important guarantee of quality to the farmer and can help to accelerate the acceptance of improved seed. For foreign seed producers, however, the certification process can be a critical obstacle to gaining access for exports. Government inspectors are not usually available in overseas locations, so the same procedures, tests and forms cannot be completed by the importer. There is always some question and a potential for dispute concerning the steps that an importer must follow to become certified. A set of standardized tests and procedures developed by the International Seed Testing Association (ISTA) exists, but many countries require additional information and more rigorous inspections. As a result, each country has somewhat different procedures and requirements, and seed producers who would export must stay abreast of the changes and differences among the certification schemes in various countries. The worst type of problem occurs when the certification agency is closely tied to the public sector seed producing organization and implements the laws unevenly between public and private domestic and foreign producers. In such a situation, domestic public production receives a cursory check, while importers are required to meet the highest standards. At best, certification is an inconvenience to trade that justifies its existence by improving seed quality in the domestic industry. At worst, it is a significant barrier to trade, a serious limitation to competition and a critical hindrance to seed industry development.

In addition to phytosanitary restrictions and certification requirements, some countries require seed trials and proof of superiority, e.g., better yield, resistance to disease, lodging, moisture utilization, or other qualities. Such trials take time, and the results can involve a level of subjectivity regarding the degree of superiority and the conditions under which the trials were conducted. Accompanying the trials or entailing separate requirements may be a cataloguing or inscription stage. This means that, separate from phytosanitary requirements, certification and perhaps even proof of superiority trials, new varieties must be included in a catalogue of approved varieties. The EC countries have catalogues and accept each other's approved varieties more readily than outside varieties. New imported varieties usually have the most difficulty in gaining access to markets, especially when inscription in an approved catalogue is required. In the case of the bacterial wilt alfalfa seed, it is likely that when disease resistant alfalfa varieties are identified, they will not be included in the EC catalogue, necessitating a period of testing and meeting other requirements even after the new phytosanitary rules are met.

Similar problems exist in packaging, labeling and documentation requirements. Seed exporters must know the rules of each country, stay abreast of the changes that occur periodically, and be able to conform to the latest required format to engage successfully in trade. These can become critical impediments for would-be seed exporters who lack experience and contacts overseas. Even established international seed traders find it costly and difficult to clear all the hurdles imposed by health and performance standards.

Economic and Structural Barriers

Economic and structural barriers to international trade in seed are typically found in Eastern Europe and Third World countries where the government has a dominant position in the production and distribution of seed. State trading agencies in seed, inconvertible currencies and barter trade are found in all Eastern European countries and many developing countries. In other countries with more of a mixture of public and private sector activity in the seed industry, the economic barriers to trade are more likely to be price controls, marketing and distributional restrictions, subsidies for domestic seed production, and discriminatory credit practices. In either case, the country's seed market is not very accessible because of the structure of the domestic economy and the pervasive influence of the public sector on seed production and seed markets.

The international transactions of centrally planned economies, such as the Eastern European countries, are all channeled through state-run entities, and there is no market for imports in the conventional sense. If trade is to be carried out, the sale must be made to the appropriate government agency and must meet the government's priorities and needs rather than the preferences of consumers or other purchasers as in a market setting. In addition, local currencies are not convertible to U.S. dollars or other types of convertible foreign exchange. This problem is also present in many developing countries because of exchange controls and other devices to direct all available foreign exchange to the government for debt servicing and other high priority uses. This means that the seed importer must not only sell the seed, but also ensure that payment is made with something besides inconvertible local currency. Even if the state trading agency wants to import seed, the currency authorities may not make the required amount of foreign exchange available to pay for the transaction. These two barriers to trade—state trading companies and inconvertible currencies—are increasingly important problems in dealing with most centrally planned and many developing countries.

A consequence of these two barriers to trade has been the emergence of barter trade in the past decade or so. In one form or another, this type of exchange has involved all the leading economies of the world. The Saudis and Americans did a large oil-for-jumbo-jets swap in the early 1980s, and lesser deals have been a regular feature of world trade in the 1980s. Seed traders, like other exporters of agricultural goods to Eastern Europe, face choices involving barter trade, buy-back transactions or some form of counter-purchase agreement. It is impossible to estimate accurately the prevalence of these activities in international seed transactions, but for world trade as a whole, estimates are in the range of 3 to 5 percent.[4] Given the existence of state trading companies and inconvertible currencies, barter could be said to facilitate trade, i.e., without barter, no trade would occur. Whether this is true, it is more an argument against state trading companies and the economic policies that have made the local currency inconvertible than an argument for barter. Barter is a distortion of optimal trading conditions in which market prices are known and several producers compete with each other to supply the market from the most efficient source. Furthermore, since barter makes it more attractive to persist with state trading companies and inconvertible currencies, it indirectly reinforces these two barriers. Barter trade will not disappear any time soon, and it must be included as a barrier to freer trade in improved seed.

The remaining types of economic barriers — price controls, marketing restrictions, subsidies for domestic producers, and discriminatory credit policies — are common among developing economies. In an effort to accelerate acceptance of improved seed among farmers, the government typically intervenes with price controls, production subsidies and/or marketing and financial support. Such policies may help in the early stages of seed industry development, but they inhibit progress in the middle and later stages. In particular, they inhibit the emergence of a vigorous private seed industry and create barriers to international trade in improved seed. The mechanisms at work are the same for imports and private domestic production. Low margins and public sector subsidies make it difficult for private firms to cover costs; consequently, privately produced seed, whether domestic or international, is unable to compete. The country's farmers lose in two ways: certain varieties of seed are physically unavailable, which restricts the options open to agricultural producers; and the lack of competition usually results in a poorer performance by the public sector than would otherwise occur. This can be seen in fewer varieties, poor quality seed, slower introduction of better seed, inattention to farmers' breeding needs, poor timing in distributing seed, or other failings. The net result is that poor performance of the seed industry retards agricultural progress.

There are various administrative mechanisms by which these policies can be put in place. Price controls can be legislated or set by the public seed producer as the market leader, and farmers become accustomed to subsidized prices. Private producers or importers then have to demonstrate a strong quality superiority and/or sell at a loss just to establish a market position. Marketing or distribution restrictions exist when the established system is based on public sector or proprietary institutions — such as the extension system or seed producing cooperatives — which are not interested in handling foreign products. Such barriers to trade are common for many products besides seed; for example, the Japanese system was characterized for many years by across the board marketing and distributional barriers to trade. In developing countries, these types of barriers have been the rule for trade in seed. Domestic credit preferences constitute a trade barrier when there is discrimination between one or more domestic seed producers and imported seed. When public sector seed is distributed through the rural credit institutions, credit policies become a barrier to trade; credit subsidies favoring one or more domestic producers inhibit competition and deprive agriculture of a wider, better range of available varieties.

Domestic seed research subsidies are a distortion of the working of the market and in the strictest interpretation could be construed as a barrier to trade in seed. However, because of the knowledge and self-production externalities present, the view held here is that they are correct policies for the development of improved seed varieties, and any losses caused by their effect on international trade are many times offset by the gains derived domestically. Anticipating a trade barrier that will be discussed below—export research subsidies—it can be correctly argued that domestic research subsidies offset similar policies abroad and merely put domestic seed producers on an equal footing with foreign producers. Although this argument is secondary to the externality consideration, they both support the recognition of domestic seed research subsidies as appropriate policies.

Political Barriers

Political barriers to trade impose embargoes on imports or exports of specific goods unless certain conditions are met. One such political barrier is the Arab boycott of firms and organizations trading with Israel. To trade with Arab countries, exporters and importers, including seed traders, must demonstrate that they meet specific political criteria. Another example of a political barrier to trade is trade in strategically important goods, such as weapons or high technology equipment. The current trend toward restrictions on the export of indigenous germplasm from countries in the zones of origin is also a political barrier to trade. For strategic or other reasons, these countries are making a political decision to embargo exports of local varieties of plants and seed that would be useful to breeders in other countries. Thus far, the embargoes do not seem to follow any pattern of favoritism, but in time this may evolve. As a means of expressing solidarity, releases may be made on a selected basis to countries with compatible political positions. Political barriers to the exchange and trade of seed have not been significant in recent decades. On the contrary, the keystone to genetic progress in developing countries, the International Agricultural Research Centers, has been a model of effective international cooperation. Unfortunately, there are signs that such cooperation may be weakening, and political barriers could become more important in the years ahead.

Export Subsidies

Most trade barriers exist in the form of policies to exclude or impede imports; however, export promotion measures such as export subsidies, dumping, credit subsidies, and tied development assistance are inefficient, market-distorting practices that constitute barriers to free and fair trade. Tied aid or grants in kind provide seed on a concessionary basis, but they require that the receiving country buy from the donor country's seed producers or accept specified amounts and types of seed. Such resource transfers may assist the recipient country, but not as much as untied assistance would. Tied assistance can severely limit the choices facing the recipient country in terms of appropriate available varieties. In addition, it reduces competition in terms of price, delivery, follow-up service, and other considerations.

Providing aid without national purchase requirements would allow the recipient country the widest possible range of choice in supplying its seed requirements. Foreign exchange and credit preferences blend into tied aid policies and have the same effects. The import is financed on a concessionary basis, e.g., below market interest rates, if the purchase is made from seed producers in the country providing the financing. Again, the choice facing the importing country is likely to be limited to fewer varieties, competition is likely to be reduced, and the ultimate decision is likely to be overly influenced by financial rather than straightforward agricultural considerations. Export production subsidies do not limit the choice of the importing country, but they do distort trading patterns in favor of subsidized producers. The result is to discriminate against more efficient unsubsidized producers — including efficient domestic producers — and shut them out of the market. An extreme type of export subsidy is predatory dumping. A foreign producer will cut prices to establish a market position and drive local producers out of business, then raise prices once the competition has been reduced. To qualify as "dumping," the price increases have to be based on monopoly market power and not product quality. It is not unusual for new entrants into a market to reduce prices in introducing their products and then raise them once the high quality of their product is recognized. This is not dumping or distortion of trade. The distortion occurs when the price cutting is aimed at driving competitors out of business and then using market control, rather than product superiority, to raise prices.

Export research subsidies, like domestic research subsidies, constitute minor barriers to trade that are justified by the significant externality resulting from the expansion of genetic knowledge. This is an example of accepting a minor barrier to trade to take advantage of the significant benefits resulting from expanded genetic knowledge. And, as was argued above, having research subsidies on both the export and domestic sides nets out most of the distortionary effect that would result from a research subsidy on only one side. Therefore, while research subsidies are, strictly speaking, barriers to trade in seed, the knowledge-intensive nature of the seed industry and the resulting externality are reasons to support research subsidies on the basis of seed market efficiency and agricultural productivity.

In addition to important natural barriers to trade and essential phytosanitary regulations, many other barriers to trade in seed have been identified. Most of these could be reduced or eliminated with minimal negative side effects. Obstacles to trade are pervasive in seed markets by comparison with manufactured goods and other agricultural inputs and products. The growth in world trade in seed is a testament to the value of the seed being traded, not to the freedom of international markets. In the next section, the consequences of trade barriers are examined.

THE EFFECTS OF TRADE BARRIERS

Depending on the nature and strength of trade barriers, they may be either partial or complete in their exclusionary impact. The difference is important but it is a matter of degree. The discussion here will focus on the exclusionary case in which foreign firms are absent from the local market because of barriers to the development of a viable domestic seed industry. The results will pertain to partial barriers to the degree these barriers shut out new seed varieties that could be used to advantage by local agriculture. The cost imposed by barriers to trade in seed, or to the entry of foreign firms as local breeders and producers, ultimately comes down to the extent of the negative effects on local agriculture. The discussion now turns to an identification of the types of losses that result from trade barriers from various points of view — the farmer, the domestic consumer, the government, the domestic seed industry, and those outside the country.

From the discussion of seed industries in both developed and developing countries, the effects of barriers to trade and distortions in domestic markets appear to be manifested in two ways: the availability of improved varieties through imports or domestic operations of international breeders; and seed production, conditioning and distribution. Inadequate performance in these areas will result in lower agricultural output and thus losses to farmers, consumers and the country as a whole. In developed countries, the barriers are usually limited to import controls and health and performance standards. Internal economic barriers are not as strong, so international firms can compete on an equal footing and make their genetic knowledge and seed industry experience available. In most centrally planned and developing economies, the internal economic distortions are much greater, and thus the contributions of private international genetic knowledge and seed industry experience are much more limited.

The exclusionary effect of trade barriers and market distortions is strongest in the developing countries. On the breeding side, the weakness or absence of private operations is partly offset by the efforts of the International Agricultural Research Centers and the national public breeding programs that utilize their research. However, it is in the production and distribution of seed that the effects of economic barriers are most pronounced. It is not so much the complete absence of improved seed that limits agricultural progress, but the inability to produce enough high quality seed and to distribute the seed to the right place at the right time. A stronger private seed industry, including both local and international firms, competing to meet farmers' needs is an essential part of the solution to this problem. Yet this will not occur as long as price controls and production subsidies to public seed producers remain as serious obstacles to private sector development. Import controls and health performance standards also create serious obstacles to agricultural progress, but they work at the margins of the system. Economic barriers are much more debilitating and undermine the core of the seed industry. For farmers, the consequences of barriers to trade are lower yield and lower incomes. The barriers work to reduce both the availability of varieties to farmers and the productivity of the seed that is available. When international breeders are absent, the pool of available germplasm is reduced. Beyond this, the weakness of private seed producers due to economic distortions tends to reduce the productivity of the varieties that are available. This occurs for several reasons: seed quality is low; the quantities of seed produced do not fit market demands; seed deliveries are late or inappropriate for local conditions; and improved seed is not well promoted and marketed. As a result, farmers' harvests are lower and their incomes thereby reduced, with rippling effects throughout the economy. Studies have calculated that, for a dollar increase in farm income due to increased production, the typical farmer spends roughly 75 percent on items such as housing, farm buildings and infrastructure, road transport, and hotel and food service. Higher incomes in these sectors benefit workers outside agriculture and lead in turn to further spending.[5] There is also a positive effect on agri-industry firms and workers who process the larger harvest or produce machinery and other inputs for farmers' use. When trade barriers and market distortions reduce the harvest, all these gains are reversed, and the incomes of farmers and many other workers suffer.

For consumers, the consequences are higher prices and reduced

quantities of food. Because harvests are down, less food is available
and the per unit cost rises. This reduces the consumer's standard of
living and limits the income available to save or spend on housing,
clothing and other goods and services. In turn, economic development
and growth for the entire country slows down. Because food is a larger
share of poor people's budgets, higher food prices hit them the hardest.
A 5 percent rise in food prices is likely to cause a 3 to 4 percent fall
in a poor person's living standard while reducing a well-to-do person's
standard by 1 percent or less. Policies that hold back agricultural output are among the most regressive kind because of the importance
of food to the poorest consumer. Other effects on the consumer include reduced nutritional levels with possible negative consequences
for work effort and productivity in the rest of the economy.

For the government, the short-term consequence of barriers to
trade and distortions in seed markets is a saving of foreign exchange
due to reduced seed imports. In the medium and long terms, however,
the effect is more likely to be a loss of both domestic revenue and
foreign exchange. Declining incomes for farmers and food consumers
imply that less will be spent on nonfood goods and services throughout
the economy, and as these are customarily the more highly taxed items,
revenues will fall. Food is usually not a highly taxed good, and consequently a switch of expenditure to food will tend to lower tax revenues.
The diminished ripple or multiplier effect will reinforce this tendency. On the international side, although reduced seed imports will result
in lower foreign exchange expenditures in the short run, over time this
will change. Depending upon the improvement in agricultural output
resulting from trade in seed and increased competition among seed
producers, imports of food will decline and/or exports will rise.
Moreover, the cost ratio of seed to food is so low that expenditure
on improved seed can repay itself in foreign exchange many times over
within a few years.

Only when a country has a surplus of food that it cannot sell will
the positive foreign exchange effect be in doubt. As a result, for
developing countries in particular, the true foreign exchange effect
of trade barriers and especially of economic distortions in the seed
sector is negative when considered over a reasonable time period. The
increase in harvests is bound to spill over into agricultural trade within
a year or two and improve the balance in international accounts by
saving imports or raising exports.

The consequences of trade barriers on the domestic seed industry
are complex and can be positive or negative, depending on the kinds
of barriers in place and the response of seed producers. Import controls will protect existing local firms, reducing competition and allowing them to charge higher prices and sell more of their current products than would otherwise be possible. The lack of competition also

reduces pressures to introduce new and better varieties. Therefore, import controls are attractive to domestic firms and in the short run are almost certain to make them more profitable. Economic barriers tend to discriminate against both foreign and local private seed producers, and thus affect negatively the private domestic industry. Similarly, export subsidies are a type of barrier or distortion that undermines the domestic industry. Over time, all trade barriers can work to weaken the performance of the domestic industry, even if they make it more profitable. This occurs when, behind protective barriers, the domestic seed industry allows itself to become inefficient and its products to become obsolete, and this is often the result of providing protection from international competition.

An example with extreme results occurred in Pakistan where local breeders had a strong position in supplying seed for rice cultivation, including some exports. New varieties developed in Malaysia were introduced and found to be much superior in yield with no significant drawbacks. The local rice seed industry arranged for the government to prohibit imports. The result was extensive illegal importation of the seed and a serious weakening of the domestic seed industry. Smuggling is by no means rare, and significant illegal importation of seed occurs in several developing countries in addition to Pakistan. The conclusion, then, is that protection from foreign competition is at best a mixed blessing. It reduces competition and raises profit margins for a while, but unless domestic seed producers meet farmers' needs, the farmers will find alternatives. The best policy for a viable seed industry is probably a progressive opening up of domestic markets, first to local production by international companies and then to imports.

The consequences of trade barriers in seed for those outside the country erecting the barriers are, in the final reckoning, rather small. International seed producers almost inevitably enjoy greatest success in their home markets, and any single foreign market is usually marginal to bottom-line profitability. Certainly, these producers are negatively impacted by being shut out of a market where their genetic knowledge and seed industry experience would enable them to earn a positive return over a number of years. However, the first few years of foreign operations usually require inflows of capital and the acceptance of losses before income exceeds costs. Even when profitability occurs, the results are not significant in terms of overall firm profitability.[6]

Likewise, foreign consumers of food are not likely to be greatly affected by trade barriers in some other country. True, the barriers reduce world food output, but unless the country is very big and the barriers very powerful, the net effect worldwide is unlikely to be noticed. There are, however, some extremely large agricultural producers and consumers whose effect on the world food balance is important.

When large countries adopt the wrong set of policies, the consequences are felt in world markets.

The Soviet Union is one example with poor policies, chronic import needs and unfortunate consequences for the world's poorest people. India, another large country, has done well in recent years, but a viable private seed industry there is still a long way from reality. The Indian seed industry relies all too heavily on economic barriers to trade and internal market distortions for seed analysts to be truly sanguine about its development over the next decade or so.

On balance, then, the effects of barriers to trade in improved seed are negative to domestic farmers, to domestic consumers and to the government—except perhaps in the very short run—and they are a mixed blessing at best for the domestic seed industry. In the early stages of seed industry development, import controls probably have a trivial effect. The markets are not there to make imports really worthwhile, so there is little distortion of trade patterns. However, when the issue turns to economic barriers and more fundamental market-distorting policies, then the debilitating effects become more serious. As the seed industry develops, import controls have a more negative effect on agriculture. Economic barriers, particularly fundamental distortions such as price controls or large-scale production subsidies, are so damaging to the seed industry that they can halt, or even reverse, progress from one stage of development to another. From this analysis, it is clear that barriers to trade in seed and distortions of the domestic seed market almost inevitably retard agricultural progress. During the first stages of agricultural and seed industry development, the effects are not significant. There are so many natural barriers to trade and natural imperfections in the market for agricultural inputs that tariffs, import controls or even economic barriers are usually not noticed. In fact, one can argue that subsidies to seed suppliers are desirable—but only if they are on the demand side, nondiscriminatory and available for all seed production, public and private, national and foreign.

The problems emerge at the next phase of agricultural and seed industry development. By this time, markets are beginning to function and policies that oppose or ignore the forces now at work are unlikely to meet with much success. As natural distortions and obstacles diminish, policy-induced distortions become more important. During the middle and later phases of seed industry development, policies must aim to reduce barriers to trade and market distortions if progress is to continue at an acceptable pace. Where policy-induced market distortions persist, seed industry development will be retarded and agricultural progress will suffer.

FOOTNOTES, CHAPTER 8

1. For a complete assessment of this, see Peter Morici and Laura L. Megna, *U.S. Economic Policies Affecting Industrial Trade: A Quantitative Assessment* (Washington, D.C.: National Planning Association, 1983).
2. Bill Chaney, Floyd Ingersoll and Wayne Underwood, Reports of ASTA/FAS Seed Mission, Brussels, Belgium (EC), May 1979 (Washington, D.C.: American Seed Trade Association, 1979), p. 2.
3. Harold D. Loden, Report of ASTA/FAS Seed Mission, Italy-Belgium-England, 1982 (Washington, D.C.: American Seed Trade Association, 1982), p. 15.
4. Gary Banks, "The Economics and Politics of Countertrade," *The World Economy* (June 1983), Vol. 6, No. 2, p. 161.
5. World Bank, *World Development Report* (WDR), 1982, p. 61.
6. There is also the matter of whether private firms properly value the knowledge that their domestic research operations generate and then make available to the overseas operations. There is no marginal cost to making this available to foreign operations, and therefore there may not be any charge for it. If this is the case, the costs of foreign operations may in fact be understated. In any event, foreign countries are obtaining knowledge at very low costs.

Policies for Seed Industry Development 9

BASES FOR FORMULATING SEED POLICIES

Every society faces fundamental choices concerning what goods and services will be produced, how resources are to be used, who is to undertake the production, for whom the goods and services are to be produced, and what exchange relationships (price ratios) will exist between different goods, services and resources. Choices must be made because, with limited resources, economic goods and services are inevitably scarce and cannot be provided free. Not all economic desires, not even all needs, can be satisfied simultaneously. Therefore, fundamental allocation choices must be made, and standards of living are constrained by resource scarcities.

Traditional societies make these choices based on custom, while more advanced societies use central planning and command decisions or a system of prices and markets. Combinations of all these may be at work simultaneously. In fact, economic choices based on custom, command and markets can be found in varying proportions in most countries today. China would incline more toward custom and central control than the United States (or Singapore), but aspects of all these systems could probably be found in both countries.

To make these choices in today's complex economies, including those of the developing countries, there are at least three strong arguments favoring reliance on a system of prices and markets. The first argument is based on theoretical considerations. Given assumptions of perfect competition, no externalities, wide availability of economic information, and so forth, it can be shown that a free market system will result in an equilibrium that will be at least a local maximization of well-being for society as a whole.[1] Although the mathematical proof that demonstrates the theory is complicated and somewhat abstract, there can be little doubt of its correctness. Make the required assumptions and a free market can be proved to be the best way to make basic economic choices. This argument is mentioned not because anyone believes that all the necessary assumptions hold in the real world—least of all in the market for seed—but because it provides a framework for analyzing the workings of the economy, identifying the important departures from the assumptions, and making informed policy choices. In the absence of a market framework, economic policymaking is inevitably ad hoc and likely to be random at best in its ultimate effects. With a market framework, it is possible to begin to analyze a problem on the basis of a system of economic knowledge, to identify the major issues and to make policy recommendations with some confidence that they will move conditions in the right direction.

The second argument is derived from experience. Economic progress over time is based on the division of labor, the accumulation of physical capital and the expansion of knowledge. As progress is made, economies tend to become more complex, and experience has shown that traditional ways of making economic choices become less and less effective. Successful advanced economies seem to have evolved toward a mixed system in which markets and prices are used to make economic choices, and political decisions work through and, in part, direct the functioning of the market process. There is inevitable tension between the market and political intervention, and different mixes are appropriate in different situations. However, experience has shown that agriculture is an economic sector where policies that work against market forces, such as administered prices and quantitative controls, have almost always failed. In agriculture, systems that depend heavily on public intervention (of which central planning is an extreme example) do not perform well. For the most part these conclusions can be generalized to seed because it is an agricultural input and, on the production side, an agricultural output. Consequently, policies aimed at advancing agricultural progress through improved seed should work in consonance with market forces to be successful.

The third argument favoring market-based economic choices is philosophical and involves a value judgment regarding individual freedom. Markets are geared to satisfy the preferences of purchasers. In a sense, individuals vote their economic preferences by the purchases they make. No other system of making economic choices gives the individual as much sovereignty; most other systems are geared to remove economic power from the individual and vest it elsewhere. This argument does not contend that an unfettered market system results in maximum freedom for all individuals, but that the basic orientation of the market system is toward satisfying the preferences of individuals. As individual freedoms are limited in an effective democratic policy, some limits on markets are desirable. The various redistributional policies at work in almost all countries attest to the need to limit some individual economic freedoms. Again, however, tension exists between the market and redistributional intervention, and experience is showing that systems characterized by excessive supression of market forces do not perform well over time, especially in agriculture.

These arguments are not intended to be definitive or to suggest that the market system is a perfect method for making economic choices. They do suggest, however, that in dealing with agriculture and seed (in any but the most traditional setting), policies that ignore the effects of market incentives or work against market forces are most unlikely to succeed.

The arguments for reliance on markets to make economic allocation choices internationally are similar to those discussed above. They are the traditional free trade arguments and, in summary, support the effectiveness of markets in:

(1) serving the interests of consumers;
(2) providing a broad range of choices;
(3) stimulating competition to produce what the market wants at the lowest possible costs;
(4) getting the most out of the available resources for society as a whole.

Once the ideas of consumer sovereignty and individual economic freedom take hold in a society, policymakers ignore these arguments at great risk. They are relevant in any but highly traditional, collectivist societies. In addition to the more general rationales for free trade and market reliance, more specific, policy-oriented arguments can be made, including:

(5) success is automatically rewarded and failure is automatically penalized;
(6) the private sector takes the risk and bears most of the cost;
(7) the strain on limited public sector resources is reduced;
(8) the public sector can concentrate on correcting for externalities and the most important market imperfections.

In this way, private individuals and organizations are principally responsible for production and giving the economic system its momentum, while the public sector intervenes selectively to offset genuine market imperfections and to undertake necessary redistribution of resources to achieve noneconomic goals.

Experience has shown that, when the public sector has taken on primary responsibility for production—especially in agriculture—the results have been almost universally unsatisfactory. Governments that have left production responsibility in private hands and worked in consonance with market forces have been much more successful in achieving economic progress. Similarly, governments that have adopted highly protectionist "inward-looking" policies by constructing an extensive system of trade barriers have had poor results. In recent decades, the countries that have shown the most dramatic progress have been "outward looking" and have made international trade an integral part of their overall plan of growth and development.

Where market imperfections and violations of the underlying

assumptions exist, the appropriate policy is one that offsets the problem directly and in a nondiscriminatory way. Often international trade can help by increasing competition. If, for example, one or two large firms or agencies control the local market, the basic premise of free and open market competition is being violated. The best solution is not to regulate the domestic producers, but to open up the economy so that competition from abroad will force the domestic producers to become more efficient and meet the requirements of the domestic market.

A less attractive solution is the creation of a countervailing imperfection, and protection is usually the worst way to do this. Nondiscrimination is very important. For instance, if the potential purchaser lacks adequate access to credit because of institutional problems, then the government may·correctly move to overcome the problem, even if it means intervening in markets. Working through existing financial institutions would be the best way, but whatever the policy, it should not discriminate for or against any producer. The purchaser should retain maximum freedom of choice. Only in this way will the imperfection be corrected without creating a new problem as a by-product.

Five general principles for public sector intervention in the workings of the market can be set forth:

(1) imperfections or market distortions should be directly addressed and removed if possible;

(2) policies should be chosen that increase competition and work through market incentives rather than restricting activity;

(3) policies should be as nondiscriminatory as possible;

(4) policies must be simple and easy to administer;

(5) intervention in markets should be selective, recognizing the administrative and other costs inherent in all such actions.

In many cases, market imperfections and distortions can be removed by the termination of an existing government program, and such action should be pursued whenever possible. In all cases, intervention should be a fallback position rather than a policy of first resort. Especially in developing countries, public sector management skills are scarce, and extensive programs of public sector intervention in, and control of, markets are almost inevitably doomed to failure. In certain situations intervention is essential, principally when significant externalities exist, and public sector resources should be husbanded for use in those cases.

The consequences of trade barriers are reduced productivity through division of labor and competition, a reduced range of purchaser choices and, in the final analysis, a reduced standard of living, as described in the previous chapter. Trade barriers do benefit those who are protected or who work for firms protected from competition.

The costs are borne by those who purchase the protected goods and services and by unprotected producers who are put at a disadvantage. Most important, the sum of the costs always exceeds the sum of the benefits. When the increase in consumer costs is estimated and compared with the number of domestic jobs created or saved, the costs are typically in the range of 3 to 10 times the wage being earned in the protected job.[2] These are rough estimates at best, but they indicate the underlying reality. Costs are higher than gains because of increasing inefficiencies resulting from distortions in market patterns. The consumer not only has to pay higher wages to protected domestic workers, but must also pay for the overall inefficiency of the domestic firm compared with cheaper foreign suppliers. In addition, the consumer also pays costs resulting from economywide distortions in both demand and production patterns.

The consequences of barriers to international trade in seed can be analyzed within the general framework discussed above. The barriers listed in Table 8–1, pp. 228–229, all distort market patterns and create inefficiencies. However, there are short- and long-term effects, and the impact of the full effects varies somewhat. Import controls, for example, while discriminating against imports, tend to help domestic seed producers, at least in the short run by reducing competition and allowing them to charge higher prices than would otherwise prevail. In the long term, however, the reduction in competition may well lead to cumulative inefficiencies and slower development of the domestic seed industry. Price controls, in contrast, hurt both domestic producers and importers.

A key distinction is the contrast between direct trade, i.e., conventional imports of commercial seed, and indirect trade, i.e., the importation of germplasm and knowledge through domestic operations of international seed companies. Import controls inhibit mainly the direct inflow of commercial seed, but allow the country to benefit from local operations of international companies that bring in improved germplasm for their local breeding program and that have considerable experience in seed production and distribution. However, economic barriers to trade, notably price controls and domestic production subsidies, discourage not just commercial seed imports, but also the establishment of local operations by international seed firms. They are the most destructive types of trade barriers and market distortions, precluding not only imports, but also local operations by international seed firms and the development of a viable domestic private seed industry.

POLICIES TO CORRECT SEED MARKET IMPERFECTIONS

Although policies should not restrict international trade flows or distort the workings of domestic markets, as argued above, exceptions occur when there are significant gaps between private and social benefits and costs. In the seed industry, the principal gap of this kind is based on the externalities related to genetic knowledge where social cost is effectively zero and social benefits are high, as discussed in Chapter 8. Because of this gap, from society's point of view, genetic research will always be underfunded in the normal workings of the market; this is true in both developed and developing countries. Another externality discussed at length earlier is based on the ease of replication of open-pollinated seed, which creates a wide gap between private costs and private and social benefits for open-pollinated seed and is the basis for plant variety protection laws. In the course of market activity, the breeder of an open-pollinated variety can never hope to receive any benefit close to the social benefit of the genetic knowledge of that variety because farmers or other seed producers can so easily replicate the variety once it has been developed.[3] Again, the replication problem relates to both developed and developing countries.

There is a third gap based on the lack of information on the value of improved seed among traditional farmers. This problem is mostly limited to developing countries in the earlier phases of seed industry development. Given the traditional farmers' incomplete information regarding the value and benefits of using improved seed, there exists a large perceived difference between private (traditional farmers') benefits and society's benefits from the wider use of improved seed.

In each instance, there is a strong argument for policies that intervene in the workings of the existing market to overcome serious imperfections. However, such policies should follow the five principles set forth above. For the knowledge externality, the appropriate policy is publicly subsidized genetic research at the international and country level, as described in previous chapters. The subsidy can take the form of publicly funded research programs, tax benefits or other kinds of subsidies to appropriate research. Such programs have been in place for several decades now, and their progress and accomplishments have ranged from very good for the IARC system and a few of the better country research programs, to adequate in most other countries. Experience has shown that the two biggest problems facing these programs are obtaining adequate funding and ensuring that the research is relevant to the needs of the country's farmers. For the replication problem, the best approach is plant variety protection, although this is by no means a fully satisfactory solution.[4] This issue is more acute in developed countries where private sector breeding programs supply the majority of varieties used in agriculture. However, as the seed industry progresses in developing countries and moves into Stages 3 and 4, similar measures may be required.

The information problem facing developing countries in the early phases of seed industry development is quite complex. Solutions must inevitably be tailored to each country's situation. Nevertheless, based on the analysis of seed industry development and the principles for intervening in markets, three suggestions are appropriate:

(1) since farmers' information is deficient, policies must have a heavy extension service aspect; this emphasis is appropriate because it targets the problem directly;

(2) subsidies or any other form of intervention to stimulate seed production should not discriminate between public and private producers;

(3) to enhance competition, the intervention policy should be based on performance rather than cost.

Subsidies for seed production are less acceptable policies. But if they are to be utilized, they should be nondiscriminatory and performance-oriented. This will maximize effective competition while lowering seed suppliers' costs. Performance-oriented means that the subsidy received by a seed producing firm or agency depends upon its level of sales and performance in meeting farmers' needs. Subsidies that are not performance-oriented but based on cost or other criterion will tend to encourage firms to meet only the criteria set for the subsidy. Cost-based subsidies do not necessarily encourage firms or agencies to meet the needs of the market, but to incur costs of the appropriate type. Personnel subsidies will encourage more hiring. Production subsidies will encourage production but not necessarily sales. In some cases, subsidized public seed operations have sold seed as fodder to reduce their unsold stocks and justify the continuation of production subsidies. One way to subsidize performance is to provide purchasers with attractive credit to purchase improved seed. The seed producer still has to satisfy the market and produce efficiently, but more farmers buy seed than would otherwise be possible.

If subsidies are to be used as a policy tool, they should be limited to a few years' duration. Subsidies are useful for moving from Stage 1 to 2 in the development of the local seed industry, but are counterproductive beyond that point. By following the above recommendations and by limiting the duration of subsidies, the worst of their negative effects can be avoided.

POLICIES FOR SEED INDUSTRY DEVELOPMENT AND GROWTH

In conclusion, a strategy will be discussed for seed industry development and expansion that explicitly includes international trade, domestic operations by foreign firms, and indigenous private sector seed firms. A good strategy must have clear objectives and learn from the experience of others. In this discussion, the objective of seed policy is taken to be the expansion of usable agricultural output, net of input costs, by supplying improved seed in the appropriate quantity and quality. Seed production, profitability, research efforts, or employment levels are not themselves objectives, but are partial indicators of whether the seed industry is growing. The ultimate objective is agricultural progress.

Experience has shown that it is in the interest of countries to have access to the widest possible range of potentially useful germplasm for plant breeding. This implies the need for extensive links with international sources of germplasm, including the IARC system, public breeding programs in other countries, and foreign private companies able to contribute to domestic agricultural progress. Domestically, seed policies should encourage public and private breeding programs, and indigenous plant breeding resources should be collected and maintained in a national seed bank. These recommendations pertain to both developed and developing countries.

Experience also has shown that there are few if any cases of government agencies doing a good job of seed production, distribution and promotion. In the developed countries with strong, viable seed industries, the public sector typically plays a minor role in seed production and distribution. Government agencies focus on basic research, applied research in small specialty crops, regulating and monitoring seed quality, and promoting improved seed through the extension system. In the more successful centrally planned economies, the state seed agencies are developing links with private Western seed firms and introducing measures to improve incentives in seed production and distribution. Many developing countries have regarded the private sector with ambivalence and have been reluctant to put it on an equal footing with public parastatal seed producing agencies. The result has been arrested seed industry development with particular weakness in production, distribution, promotion, and follow-up services. Experience has shown that a mixed strategy combining public and private sectors seems to work best.

The private sector should be encouraged to assume as many functions as possible to conserve scarce public financial and administrative resources. The public sector has to take action when significant externalities and market imperfections exist, but it should be cautious about expanding its involvement beyond the policies described above. Experience has shown that, except when justified by significant natural market imperfections or externalities, public sector intervention in seed markets is likely to have negative results. However, within the seed industry and in other industries related to agriculture, there are areas where it is critical that the public sector intervene. It is far better to focus public sector resources where activist policy is absolutely essential and to ensure that policies in these areas achieve the needed results.

Direct trade is important only in the later stages of seed industry development and then only when foreign varieties are appropriate. The greatest potential for trade exists between countries in the same latitudes and at the same phase of agricultural and seed industry development. Such countries are similar in terms of crops grown, appropriate varieties and seed requirements of the farmers. As variations in these factors increase, the natural potential for direct trade declines. In the early stages of seed industry and agricultural development, particularly when traditional agriculture predominates, there is little potential for direct trade. Natural barriers are always a factor, and local market imperfections are overwhelming in traditional agriculture. Direct seed trade between developed and developing countries is limited by the latitude difference for forage and grain crops. When direct trade does occur with developing countries, it frequently involves parastatal organizations, is subject to extensive import controls, and does not really represent the underlying state of market demand. Over time, as seed industry progress is made, the potential for direct trade tends to increase. Countries usually benefit from trade in seed in terms of greater choice, higher yields and better national standards of living. Insofar as natural barriers do not rule out direct trade, policies ought to reduce trade barriers progressively as the seed industry develops.

Indirect trade in the form of international firms setting up domestic operations in a country is probably even more important than direct trade. Typically, these firms introduce new varieties into the local market and bring with them considerable experience in production, distribution and promotion operations in which public sector agencies are weak. Ideally, a host government can convince a few international companies to set up breeding operations to cross international varieties with locally adapted varieties, resulting in high yielding varieties suitable for local agriculture. For developing countries that cannot usually use varieties bred for the different latitudes of the developed countries, such locally oriented breeding programs are the best means of overcoming natural barriers to direct trade and gaining access to the widest possible range of germplasm.

Local operations by international seed firms can assist seed industry development at all stages. Realistically, however, it is unlikely that any firm will enter a market until it has moved out of Stage 1 — the traditional agricultural phase. Once it is clear that parts of the agricultural economy are in Stage 2 and demand for improved seed is strong, it is realistic for seed policymakers to anticipate genuine interest from established international firms. Probably the best way to convert interest into domestic operations and a long-term commitment to a country's agricultural future is to follow the five principles for public sector market intervention outlined above. Seed policy has to be committed to fostering a viable private sector within the industry and has to be realistic about the policies needed to achieve it. In particular, the need for profitability is self-evident. In the face of price controls or a subsidized public sector seed supplier, tax incentives would be ineffective because there are no profits to begin with. The best policy is a minimalist public sector that takes on only seed activities which the private sector cannot do and has a long-run policy of spinning off responsibilities to the private sector as soon as possible.

The importance of indirect trade cannot be underestimated, even for advanced seed industries with extensive direct trade flows. It was noted in Chapter 5 that in France hybrid corn seed producers include all-French firms and firms that draw on American breeding materials as well as French varieties. Some of the best selling and highest yielding hybrid varieties in France are based on recently introduced international germplasm. There is no doubt that French farmers have benefited from the diversity and range of varieties available to them. All other advanced seed producing countries, including the United States, have also benefited from the internationalization of breeding. This benefit does not come from exotic germplasm, which is not of great consequence in the leading crops in developed countries, but from the diverse breeding lines that international companies can use to develop new commercial varieties. If the most advanced seed producing countries in the world can benefit from local operations by international firms, it follows that less advanced countries can as well. In fact, smaller countries can be expected to benefit even more because they are less able to finance and staff several optimum-scale seed breeding and producing operations to compete with each other. Smaller countries more than larger countries need to draw on international genetic and technical knowledge to have the necessary minimum competition among seed producers.

In the coming years, developing countries should be able to benefit from greater contact with each other. Currently, there is a fairly extensive exchange of germplasm, facilitated by the IARC system, but there is enormous potential for increased direct imports from other developing countries at similar latitudes. This should be especially important for small countries that cannot utilize most of the varieties bred for developed countries and are not large enough markets to attract international firms for local breeding programs. For those countries, the direct importation of appropriate varieties from larger developing countries, such as Brazil, Mexico, India, or the Philippines, is perhaps the best means of expanding the range of locally available varieties and increasing competition in the local seed market. Once the seed industries in the leading developing countries enter Stage 3 or 4 and the private sector becomes viable, those countries ought to be able to establish export markets in other countries with similar agricultural and ecological conditions. Countries like Brazil have enormous export potential in seed and national interest in the reduction of seed trade barriers around the world. At the moment, this interest is latent, but as they become more capable and more competitive, it will begin to emerge.

The successful development of a country's seed industry requires that international trade and technical expertise be incorporated explicitly into policy. First, countries should learn from international experience. Policies that have failed elsewhere should be avoided; those that have succeeded elsewhere should be adapted to local conditions and implemented. Second, international firms should be included explicitly in seed industry development strategy. International firms can start to contribute in Stage 2 and will greatly facilitate progress beyond this phase. If necessary, the desired firms should be approached to see if they are willing and able to participate in the development of the local seed industry. Seed policy should aim to get as much gain as possible from the participating international seed companies by encouraging a long-term, mutually beneficial relationship. Third, barriers to direct trade ought to be reduced progressively over time. By Stage 3, the domestic seed industry should be ready to begin facing outside competition. Progress to Stage 4 and eventual exports require that the local industry be competitive internationally in supplying seed for the domestic market and comparable markets abroad. By the time a country has reached Stage 4, there should be a stable, continuing, market-determined flow of seed imports and seed exports. Trade barriers of all kinds should be minimal, and the regulatory role of the government should be limited to necessary phytosanitary and quality control standards.

International trade, both direct and indirect, has thus far been underutilized in national seed strategies. The situation is somewhat better among developed countries, but even there, international trade is usually underdeveloped relative to its potential. Increased trade in improved seed and the agricultural gains that can result are an unheralded and underemphasized policy objective that can help to improve the world food balance over the next decade. There are many benefits of trade, but they are, fundamentally, that farmers have a wider choice of improved seed and that increased competition forces the local industry to improve in the supplying of seed required to achieve agricultural progress. One of the greatest weaknesses of seed development strategies has been the absence of explicit goals for utilizing international genetic and technical knowledge. The existence of the IARC system has helped offset this absence at the breeding end of seed activities. However, the creation of trade barriers—especially economic barriers—and the implicit rejection of the benefits of direct and indirect international trade have contributed substantially to poor performance at the production, distribution and promotion end of operations. Greater international exchange of quantities of seed and knowledge of seed is an unrecognized advantage that must be utilized in the years ahead if world agricultural progress is to continue at acceptable rates.

FOOTNOTES, CHAPTER 9

1. See, for example, Kenneth J. Arrow and Gerard Debreu, "Existence of an Equilibrium for a Comprehensive Economy," *Econometrica* (July 1954), Vol. 22, pp. 265–290.
2. See, for example, Hobart Rowen, "Get Rid of Auto Quotas," *Washington Post*, November 4, 1984, p. 21.
3. Note that the social benefit is almost assuredly higher than the private. The marginal relationships are: $0 \approx$ social cost $<$ private cost $<$ private benefit $<$ social benefit, whereas for perfectly competitive markets these would all be equal to the price of the good. For hybrid varieties, the gap between private cost and private benefit is much less, but social costs and benefits still diverge significantly.
4. See Chapter 2 for more discussion of both these externalities.

NPA Board of Trustees and Officers

DATE DUE

SB114
A3
M3
1987x